Lamont Stilwell

Brief Lessons on the Human Body

Lamont Stilwell

Brief Lessons on the Human Body

ISBN/EAN: 9783337371517

Printed in Europe, USA, Canada, Australia, Japan

Cover: Foto ©berggeist007 / pixelio.de

More available books at **www.hansebooks.com**

BRIEF LESSONS

ON

THE HUMAN BODY.

AN ELEMENTARY TEXT-BOOK ON ANATOMY, PHYSIOLOGY, HYGIENE, AND THE EFFECTS OF STIMULANTS AND NARCOTICS ON THE HUMAN SYSTEM, PREPARED FOR THE USE OF INTERMEDIATE AND GRAMMAR SCHOOLS.

BY

LAMONT STILWELL.

NEW YORK:
W. D. KERR, PUBLISHER,
16 Astor Place
1889.

PREFACE.

THIS book is the result of several years' practical work in the class-room. It has been the practice of the author to teach the subject of Physiology topically, objectively, and with such simple experiments as can be performed with inexpensive apparatus. The notes prepared for the use of pupils have been revised and extended from time to time, until they have grown into the substance of this little book.

In its preparation a logical order has been observed in the arrangement of subjects and in the sequence of topics, so that each lesson may be presented to the class orally and objectively before being assigned from the text. The subject of each topic has been placed in the form of a question. Each question is a useful and interesting question, and deals with some important fact in regard to the structure or function of some of the organs of the body.

In answering these questions it has been assumed that every teacher who uses this book will supplement each lesson with interesting conversation, and, as far as possible, with object-teaching and experiments. Consequently all matter which should be introduced orally by the teacher, or which should be brought out by observation, questions or experiments before the class, has been purposely omitted. An effort has been made to have the answers clear and concise, within the understanding of children, and as free as possible from scientific terms.

Special care has been taken to present the leading facts of Anatomy and Physiology in such a way as to show clearly the reasons why the health of the body requires us to follow or avoid certain courses of conduct. In this way the general laws of health will be fixed in the minds of pupils as guiding principles, from which they can draw their own conclusions regarding special cases.

In preparing this work the author has examined a great many text-books and other authorities upon the subject, and has used what he considered the best material, wherever found. A list of the text-books consulted, with the names and addresses of the publishers, is given on the following page. L. S.

East Orange. N. J.
September 1, 1889.

TEXT-BOOKS CONSULTED.

WALKER'S Anatomy, Physiology and Hygiene; published by Bacon & Allyn, Boston, Mass. Hitchcock's Anatomy and Physiology, and Smith's Elementary Physiology and Hygiené; published by Ivison, Blakeman, Taylor & Co., New York. Cutter's Analytic Anatomy, Physiology and Hygiene; published by J. B. Lippincott & Co., Philadelphia. The Human Body, by H. Newell Martin; published by Henry Holt & Co., New York. Steele's Hygienic Physiology; published by A. S. Barnes & Co., New York. Hutchinson's Physiology and Hygiene; published by Clark & Maynard, New York. Dunglison's Physiology; published by Porter & Coates, Philadelphia. Kellogg's First Lessons in Physiology and Hygiene; published by Harper & Brothers, New York. Brands' Lessons on the Human Body; published by Leach, Shewell & Sanborn, New York. Blaisdell's Young Folks' Physiology; published by Cowperthwait & Co., Philadelphia. Huxley and Youmans' Physiology and Hygiene; published by D. Appleton & Co., New York. Stowell's Syllabus of Lectures; published by C. W. Bardeen, Syracuse, N. Y.

CONTENTS.

CHAPTER		PAGE
I.	A Talk to Pupils about our Bodies and the Object of this Book	9
II.	A General View of the Parts and Organs of the Body	12
III.	The Bones, or the Framework of the Body	15
IV.	Classes of Bones—Bones of the Head and Trunk	21
V.	Bones of the Upper and Lower Extremities	24
VI.	Hygiene of the Bones and Teeth	27
VII.	Food and Drink	32
VIII.	Organs of Digestion	37
IX.	Organs of Digestion—*Continued*	43
X.	Hygiene of Organs of Digestion	47
XI.	The Blood and its Circulation	55
XII.	The Organs of Circulation	59
XIII.	Hygiene of Organs of Circulation	66
XIV.	Organs of Respiration	73
XV.	Respiration and Voice	77
XVI.	Hygiene of Respiration	81
XVII.	Organs of Motion—Muscles	86
XVIII.	Hygiene of Muscles	90
XIX.	The Nervous System	95
XX.	Hygiene of the Nervous System	100
XXI.	The Skin and Organs of Special Sense	105
XXII.	The Skin and Organs of Special Sense—*Continued*	108
XXIII.	Hygiene of the Skin and Organs of Special Sense	114
XXIV.	Stimulants and Narcotics	118
XXV.	Stimulants and Narcotics—*Continued*	123

CHAPTER I.

A TALK TO PUPILS ABOUT OUR BODIES AND THE OBJECT OF THIS BOOK.

THE great world about us has in it many curious and interesting things. If we stop to look, we find them everywhere. Many of them are so close to us, and have become so familiar, that unless we stop to study them, they fail to excite our wonder or admiration. Among plants, animals, or anything else in all nature, there is nothing more wonderful than our own bodies. No study is more pleasing, no knowledge is more valuable, than that so closely connected with ourselves.

Boys and girls find themselves taller and larger this year than they were last! How did they become so? "They grew." Yes, they ate and grew; but how did the food which they ate find its way to the very extremities of the body, and become flesh and bone? As you have been out in the fields alone on a cold winter's day, did you never ask yourself why it was that while everything about you was cold, your body was warm? Our homes, the horse-cars, and the schoolroom must have fires in them to keep them warm; but what is the source of the heat which keeps our bodies warm? It would be of interest to know. Again, we notice that unless somebody moves them, the trees live and spend their lives in the places where they began to grow; while we and other animals move about where we will. Both, we say, are alive; but how is it that we have a power which the tree has not? A clock ticks, and its wheels move. I ask you why, and you say that it has a spring in it which, when wound up, gives to the wheels the power of motion. When it runs down the wheels stop, and somebody must wind it again. Our hands move, but they do not wait like the boughs of the tree for a breath of wind to stir them. Our bodies move, but they do not have to be wound up like the clock. Why is it? Where do we get the power? Why must we eat and drink? Why must we breathe? How do we see and hear? Why must we sleep? Why is it that improper food or exposure will make us sick? By and by, on account of sickness, an accident, or old age, we become as cold and as motionless as the stones themselves. What has happened?

These are only a few of the great many interesting questions which we can ask about our bodies. Some of them neither you, nor I, nor any one else, can answer fully; but many of them can be answered, and perhaps all of them in part.

To answer some of these questions is the object of this little book. We hope that some of the boys and girls who use it will become so much interested in this study that they will be eager to study other books and to make investigations for themselves. Many things can be best learned by handling and examining them instead of reading about them. We cannot see how things are arranged, or what takes place within our own bodies; but, if we will, we can find out much by carefully examining the dead bodies of animals. A dead rabbit, the heart of a sheep, or the bones of an ox, will afford a more interesting lesson than any book.

As we study, we shall find that in some respects our bodies are like houses; they afford a home and protection to what is within them. In other respects they are like machines, each part of which has some particular work to do. If we wish to know all about a machine, a pump, a watch, or a sewing-machine, we study all its parts, how they are put together, and what each part does. In order that the machine shall work well and last long, we must know how to use it and how to take care of it. It is just so with the machine which we call our body. You know it sometimes gets out of order, and we say we are sick, and we send for the physician in order that he may advise us what to do to remove the difficulty, and to put the machine in good working order again.

To avoid sickness and pain, and to prolong life and strength as long as possible, is a matter of great importance to each one of us. In order to do this, we must understand the human machine, our body. We must learn all its parts, know how they are put together, what they are for, and how they work. We also must learn how to take care of the body, so that it will be strong and last as many years as possible, and give us the least amount of pain.

Each part of the body which does some work is called an *organ*. The eye, the ear, the hands, the heart, the stomach, etc., are all organs. The study of the organs of the body, how they are made and put together, is called *Anatomy*. The study of the office of each one of these organs, or what it does, is called *Physiology*. The study by which we learn to take care of the organs of the body is called *Hygiene*. This book is intended to teach you something of Anatomy, Physiology, and Hygiene.

CHAPTER II.

A GENERAL VIEW OF THE PARTS AND ORGANS OF THE BODY.

What are the principal parts of the body?

The principal parts of the body are the *trunk;* the *head;* the *upper extremities*, or arms; and the *lower extremities,* or legs.

What are the chief parts of the trunk?

The chief parts of the trunk are the *chest,* the *abdomen* (ab-dō'-men), and the *backbone.*

What are the chief parts of the head?

The chief parts of the head are the *skull* and *face.* The *forehead, temples, cheeks, eyes, ears, nose, mouth,* etc., may be called parts of the face.

What are the chief parts of the upper extremities?

The chief parts of the upper extremities are in each arm a *shoulder,* an *upper-arm,* a *fore-arm,* a *wrist,* and a *hand.* The *fingers* are parts of the hand.

What are the chief parts of the lower extremities?

The chief parts of the lower extremities are in each leg a *hip,* a *thigh,* a *lower-leg,* an *ankle,* and a *foot.* The *toes* are a part of the foot.

With what is the entire body covered?

The entire body is covered with the *skin*, to which belong the *hair* of the head, and the *nails* which grow at the ends of the fingers and toes.

What is an organ of the body?
An organ is any part of the body which has some special work to do.

Name the principal sets of organs in the body.
The principal sets of organs are the *bones*, which give form and support to the body; the organs by which we digest our food, called the *organs of digestion;* the organs which pump and carry the blood through the body, called the *organs of circulation;* the organs used in breathing and talking, called the *organs of breathing;* the organs by which we know, see, hear, feel, etc., called the *organs of the nervous system;* the organs which enable us to move any part of the body, called the *muscles*, or the *organs of motion.*

What is the bony framework of the body called?
The bony framework of the body is called the *skeleton.*

What is the principal organ of digestion, and where is it situated?
The principal organ of digestion is the *stomach*. It is situated in the upper part of the abdomen.

What is the principal organ of circulation, and where is it situated?
The principal organ of circulation is the *heart*. It is situated in the central part of the chest.

What are the principal organs of breathing, and where are they situated?
The principal organs of breathing are the *lungs*. They are situated in the chest, one on each side.

What is the principal organ of the nervous system, and where is it situated?
The principal organ of the nervous system is the *brain*. It is situated in the cavity of the skull.

What are the muscles, and where are they situated?

The muscles are the reddish flesh, or lean portions of the body. They are situated in all parts of the body.

What are all the solid parts of the body called?

All the solid parts of the body, such as bones, muscles, fat, etc., are called *tissues*.

CHAPTER III.

THE BONES, OR THE FRAMEWORK OF THE BODY.

What are the uses of the bones?

The uses of the bones are, to protect the delicate organs which are enclosed by them; to support the body and to give it form; and to act as levers by which the muscles attached to them may move the body.

Name some important organs which are protected by the bones.

The brain is protected by the bones of the skull. The heart and lungs are protected by the bones which enclose the chest. The spinal cord is protected by the bones of the spinal column, or backbone. The eye is protected by the bones of the face.

What can be said about the form of the bones?

The bones vary greatly in form. Some are *long*, some are *short*, some are *flat*, and others are irregular in form.

Mention some bones of the body which may be called long bones.

The bones of the upper-arm, fore-arm, thigh, and lower-leg, may be called long bones.

Mention some bones which may be called short bones.

Examples of short bones are found in the wrist, ankle, and spinal column.

Mention some examples of flat bones.

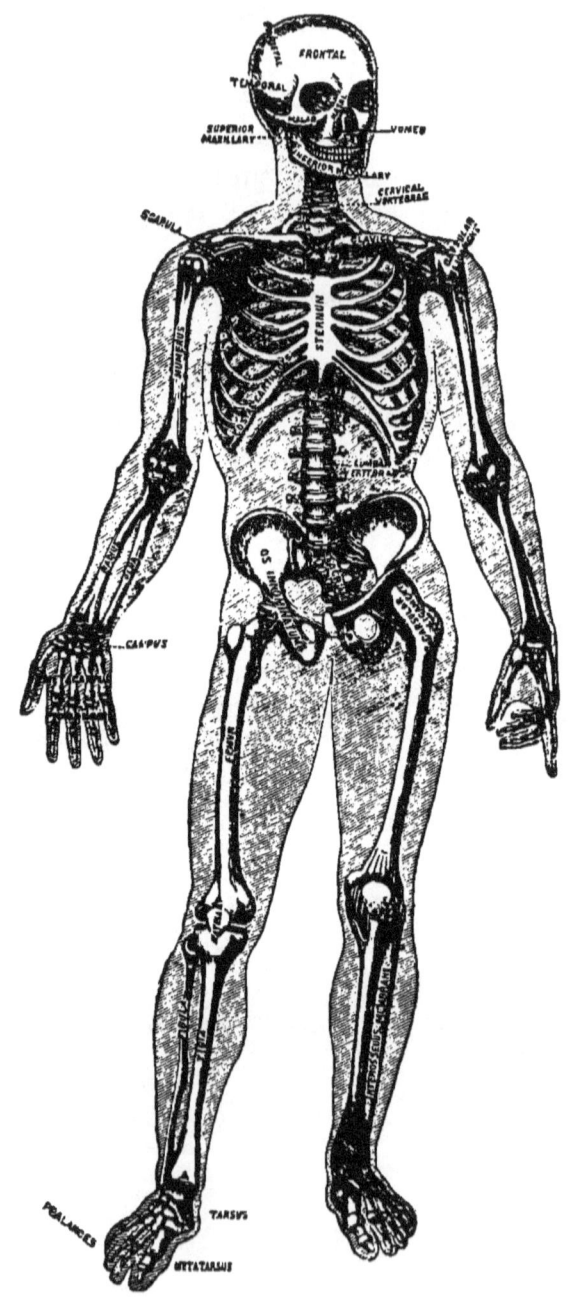

THE SKELETON.

The Head, 28 Bones.

Occipital (*base of skull*)	. .	1	Palate (*back part of roof of mouth*)	2
Parietal (*sides of skull*)	. .	2		
Temporal (*temples*)	. . . .	2	Lachrymal (*in orbit*) . . .	2
Frontal (*forehead*)	. . . .	1	Malar (*cheek-bones*)	2
Sphenoid (*behind the face*)	. .	1	Superior Maxillary (*upper jaw*)	2
Ethmoid (*behind the face*)	. .	1		
Nasal (*bridge of nose*)	. . .	2	Inferior Maxillary (*lower jaw*)	1
Vomer (*between nasal fossæ*)	.	1		
Turbinated (*on walls of nasal fossæ*)	. . . .	2	Malleus (*in the ear*) } very small {	2
			Incus (*in the ear*)	2
			Stapes (*in the ear*)	2

The Trunk, 52 Bones.

Vertebræ }		. 24	Hyoid (*in the neck*)	1
Sacrum } (*backbone*)		. 1	Ribs	24
Coccyx }		. 1	Sternum (*breast-bone*) . . .	1

The Upper Limbs, 64 Bones.

Scapula (*shoulder-blade*)	. .	2	Ulna (*fore-arm*)	2
Clavicle (*collar-bone*)	. . .	2	Carpus (*wrist*)	16
Humerus (*arm-bone*)	. . . .	2	Metacarpus (*hand*)	10
Radius (*fore-arm*)	. . . .	2	Phalanges (*fingers*)	28

The Lower Limbs, 62 Bones.

Innominate (*hip-bone*)	. . .	2	Fibula (*leg*)	2
Femur (*thigh-bone*)	. . . .	2	Tarsus (*ankle, heel, instep*)	. 14
Patella (*knee-pan*)	. . . .	2	Metatarsus (*flat of foot*) . .	10
Tibia (*leg*)		2	Phalanges (*toes*)	28

Examples of flat bones are the bones of the skull, the ribs, the breast-bone, and the shoulder-blades.

Why are some bones hollow?

If the bones of the limbs were solid, they would have to be much heavier. Their hollow form gives them greater strength than the same amount of bone would have in a solid form.

Describe the structure of a bone.

The outside part of a bone is hard, but within it is somewhat softer and hollow. The hollow portions are filled with a spongy substance called the *marrow*.

Of what materials are bones composed?

Bones are composed of *animal matter*, a jelly-like substance, and of *mineral matter*, mostly lime.

What is the use of the animal matter?

The use of the animal matter is to give toughness and elasticity to the bones.

What is the use of the mineral matter?

The use of the mineral matter is to give hardness and stiffness to the bones.

How do the bones of an infant differ from those of an old person?

The bones of an infant are merely cartilage; they contain an abundance of animal matter, but less mineral matter; they are tough and not easily broken. The bones of an old person contain more mineral matter, which makes them hard and brittle, and more easily broken.

How do bones grow?

The materials of which bones are composed are continually being worn out and replaced by new material. Bones grow, chiefly, by additions to their free ends and surfaces. The blood circulates freely through the bones, and supplies them with the materials which are required for their growth and nourishment.

How is a broken bone repaired?

The blood carries and leaves at the broken place a watery fluid which contains the materials of which gristle is formed. This hardens and cements the broken ends together until the blood in the same way gradually adds the mineral matter and completes the union.

How are the bones held together?

The bones are held together by strong bands of tissue, called *ligaments*. The muscles also help to hold them in place.

What is a joint?

A joint is the place where two bones unite.

How are joints classified?

Joints are classified as *movable, immovable,* and *mixed.*

Give examples of each kind.

The joints of the bones of the arms and legs are movable. The joints of the bones of the skull are immovable. The joints of the bones of the spinal column (backbone) are mixed.

How many kinds of movable joints are there?

There are three kinds of movable joints. They are called *hinge joints, ball-and-socket joints,* and *irregular joints.*

Give examples of each kind.

The joints of the elbows, knees, and fingers are hinge joints. The joints of the shoulders and hips are ball-and-socket joints. The joints of the wrists, ankles and neck are irregular joints.

How do these joints differ from each other?

The hinge joint permits motion in only two directions, as

backward and forward, like the hinges of a door. The ball-and-socket joint allows motion in every direction, forward, backward, and in a circular manner; it consists of a ball at the end of one bone, which fits into a socket in the end of another. The irregular joint is one in which the surface of one bone moves only a little on the surface of another; such joints are useful where little motion but great strength is required.

How does nature provide for the oiling of the joints?

In the cavity of each joint there is a thin lining, or membrane, which gives out a fluid called joint-water. This serves as oil to the joint.

CHAPTER IV.

CLASSES OF BONES.—BONES OF THE HEAD AND TRUNK.

How many bones in the human skeleton?
The number of bones in the human skeleton is 206. The number varies somewhat at different periods of life. That which is merely gristle, or cartilage, in infancy, becomes bone later in life.

How may the bones of the skeleton be divided?
The bones of the skeleton may be divided into four classes, viz.: bones of the *head*, bones of the *trunk*, bones of the *upper extremities*, and bones of the *lower extremities*.

How many bones in the head?
In the head there are twenty-eight bones. Eight of them form the skull, fourteen are in the face, and six in the ears. (For the names of the bones of the head see table on page 17.)

Describe the construction of the skull.
The bones of the skull form a cavity in which the brain is situated. These bones are united by peculiar notched joints, called *sutures*. The cavity has an opening at the bottom through which the spinal cord enters. There are no other openings except very small ones through which nerves and blood-vessels enter or pass out. The upper bones of the skull consist of two hard plates, with a spongy layer between them.

Describe the uses of the bones of the head.
The bones of the skull and face protect the brain and

the organs of sense—smell, taste, hearing, and sight—from injury. The bones of the ear are an aid in hearing. The jaws contain the teeth, and are used in masticating food.

What is the only movable bone in the head?

The only movable bone in the head is the lower jaw. It can be moved in five different directions. Its joints are very liable to dislocation.

How many teeth appear in infancy? In later life?

The temporary, or *milk* teeth, which appear during infancy, are twenty in number, ten in each jaw. The *permanent* teeth, which appear later, are thirty-two in number, sixteen in each jaw.

What names are given to the permanent teeth, and how many of each kind?

The names and number of the permanent teeth on each jaw, beginning at the back part of the mouth, are: two wisdom, four molars, four bicuspids, two canine, and four incisors.

Are the teeth regarded as bones? Why?

The teeth are not regarded as bones. They differ from bones in several respects. They are covered with an enamel, which bones have not; when broken a tooth decays, a bone will repair itself; teeth differ from bones in manner and time of growth; teeth do not appear in the infant at birth, bones do; the first set of teeth drop out and are replaced by others, no such thing takes place with the bones.

How many bones are found in the trunk?

In the trunk there are fifty-two bones. They are located as follows: in the spine, twenty-four bones; ribs, twenty-four bones; breast-bone, one; sacrum, one; coccyx, one; at base of the tongue, one. (For the names of the bones of the trunk, see table, page 17.)

What cavities are formed by the bones of the trunk?
Two cavities are formed by the bones of the trunk. The upper one, called the *chest*, is enclosed by the spine, ribs, and sternum; the lower one, called the *abdomen*, is enclosed by the spine at the back, the bones of the pelvis below, and by muscles in front.

What organs are found in the chest?
The lungs and heart are found in the chest.

What organs are found in the abdomen?
The abdomen contains the stomach, intestines, liver, kidneys, and other organs. It is the largest cavity of the body.

Describe the spinal column.
The spinal column extends from the head to the bones of the pelvis. It consists of twenty-four bones so put together that the body can turn on them readily. These bones are called the *vertebræ*. Through the spinal column there extends a tube-like canal, called the *spinal canal*, which contains the spinal cord.

What is the use of the spinal column?
The spinal column protects the spinal cord; it holds the body in an upright position, yet allows it to bend when we will; it supports the weight of the head, arms, chest, etc.

Describe the ribs.
The ribs are slender, curved bones, which help to enclose the cavity of the chest. They are arranged in pairs, twelve on each side.

What is the use of the ribs?
The use of the ribs is to afford protection to the lungs, heart, and great blood-vessels. They also furnish an attachment for the muscles which move the walls of the chest when we breathe.

CHAPTER V.

BONES OF THE UPPER AND LOWER EXTREMITIES.

How many bones in the upper limbs, or arms?

In the upper limbs, or arms, there are sixty-four bones. (For their names see table, page 17.)

Describe the scapula, or shoulder-blade.

The scapula, or shoulder-blade, is a broad, thin, flat, triangular bone situated at the top and back of the chest. It is not directly attached to the trunk, but is held in place by muscles. It is connected with the collar-bone, and contains the *socket* of the shoulder-joint.

What is the use of the scapula?

The use of the scapula is to make an attachment for the muscles of the shoulder, and to connect the arm with the trunk of the body.

Describe the clavicle, or collar-bone.

The clavicle, or collar-bone, is a long, slender bone which extends from the breast-bone and first rib to the shoulder-blade.

What is the use of the clavicle?

The use of the clavicle, or collar-bone, is to brace the shoulders back and to hold them out from the chest.

Describe the bones of the arm.

That part of the arm between the shoulder and elbow consists of a single bone, called the *humerus*. That part between the elbow and wrist consists of two bones, called the *ulna* and the *radius*. To these bones are attached a large number of muscles which move the hand and fingers. The radius is on the thumb side of the arm, the ulna is on the side of the little finger.

What advantage comes from having two bones in the forearm?

It permits an arrangement of joints which allows the fore-arm and hand to take a rolling motion which turns the palm of the hand up or down.

Describe the wrist, or carpus.

The wrist is situated between the fore-arm and the hand. It contains eight bones, irregular in shape, arranged in two rows. The bones are held together by strong ligaments, and form an *irregular* joint.

Describe the bones of the hand.

The bones of the palm of the hand are five in number in each hand. They are joined at one end with the bones of the wrist, at the other with the bones of the fingers or the thumb. There are three bones in each finger, but only two in each thumb; they are called the *phalanges*.

How many bones in the lower extremities, or legs?

In the lower extremities, or legs, there are sixty-two bones. (For the names of these bones, see table, page 17.)

What is the pelvis, and what is its use?

The pelvis is the bony structure at the base of the trunk. Its use is to make a foundation for the support of the spinal column and for the weight of the body above it. The hip-bones, which form a part of the pelvis, contain the sockets for the attachment of the thigh-bones, and thus connect the trunk of the body with the lower extremities.

Describe the hip-joint.

The hip-joint is a ball-and-socket joint. At the upper

end of the thigh-bone, or *femur*, there is a ball which fits snugly into a deep socket in the side of the pelvis. The hip-joint is similar to the shoulder-joint, and permits motion in every direction.

Describe the knee-joint.

The knee-joint is a hinge joint. It is formed by the joining of the thigh-bone with one of the bones of the lower leg, the *tibia*. Over the joint in front is a little heart-shaped bone, called the knee-pan, or *patella*. It protects and strengthens the joint.

Describe the bones of the lower leg.

The lower leg has two bones, the *tibia* and *fibula*. The tibia is a strong bone. It has a sharp ridge in front, called the shin. The fibula is the outer bone of the lower leg. It is fastened at both ends to the tibia. It serves as a brace to the tibia, protects the ankle-joint, and affords an attachment for muscles.

Describe the ankle.

The joint at the ankle is similar to that of the wrist. The bones here are seven in number, and form the ankle, heel, and instep.

Describe the bones of the foot.

The bony structure of the foot is quite similar to that of the hand. Besides the seven bones of the ankle and instep, which have just been mentioned, there are in each foot nineteen other bones,—in the "flat" of the foot, five bones, called the *metatarsals;* in the toes, fourteen bones, called the *phalanges*,—making twenty-six bones in all.

CHAPTER VI.

HYGIENE OF THE BONES AND TEETH.

From what sources does the body derive the materials for making bones?

The materials for making bones are derived from the food we eat and the water we drink.

In order that the bones may be strong and healthy, what must our food contain?

Our food must contain a proper amount of bone-making material, otherwise our bones will become weak and unhealthy.

What disease of the bones is sometimes caused by want of proper food?

The disease called *rickets* is caused by a deficiency of mineral matter in the bones. The bones become soft, and are drawn out of shape by the muscles, producing deformity.

Mention some articles of food which are rich in bone-making material.

Milk, eggs, oatmeal, and bread made from coarse flour or whole grain, are foods which are rich in bone-making material. In making fine white flour the miller takes out the part that is most useful for making strong and healthy bones.

Name some foods which are deficient in bone-making material.

Sugar, candy, fat meats, and bread and cakes made from fine flour, are deficient in bone-making material.

What is the cause of " bow-legs" in infants?

The bones of infants are soft and easily bent. If young children are allowed to stand on their feet too much, the weight of the body bends the legs outward. As the bones gradually harden they retain their curved shape and become permanently deformed.

What effect has exercise upon the bones?

As is the case with all other organs of the body, a proper amount of exercise helps to keep the bones in a strong and healthy condition. The exercise should not be too severe or too long continued.

What deformities are caused by improper positions in sitting, standing, or working?

The bones of young persons, being soft and flexible, are easily bent out of shape by improper positions in sitting, standing, or working. In this way hollow chests, round shoulders, curved spines, and other deformities are formed.

What injurious effects may come from tight clothing about the waist?

Serious distortion of the spine and bones of the chest may be caused by tight clothing about the waist. If worn by young people, the ribs are pressed inward so that they interfere with the proper action of the important organs situated in the cavity of the chest: the heart, lungs, stomach, and liver. Diseases of the liver, dyspepsia, and consumption are often the results.

What is the disease called a " felon"?

The disease called a *felon* is an inflammation that commences in or beneath the *periosteum*, a membrane which covers the bone. It is very painful. Relief is usually found by opening through the periosteum to the surface of the bone. The quicker the incision is made, the less risk and the less suffering.

Why is it a bad practice for children to "crack" the knuckles by pulling the fingers?

The practice which children sometimes have of pulling the fingers so as to "crack" the knuckles is foolish and harmful. It weakens the joints, and causes them to grow large and unsightly.

Is the use of alcohol and tobacco injurious to the bones?

The use of alcohol and tobacco does not have so marked an effect upon the bones as upon some other organs. A man's height depends upon the length of his bones. The boy who smokes cigarettes, or who uses alcoholic liquors, is likely to be so stunted that even his bones will not grow to a proper length, and he will become dwarfed or deformed.

Mention three things to be observed in the care of the teeth.

The teeth should be kept clean; they should not be used to crush hard substances; teeth which have begun to decay should be filled or removed.

Why should the teeth be kept clean?

If particles of food are left between the teeth, on account of the moisture and warmth of the mouth they begin to putrefy. This not only causes the teeth to decay, but makes the breath very offensive.

How can the teeth best be kept clean?

The teeth are usually best kept clean by the use of a soft tooth-brush and pure water thoroughly applied at least once in the twenty-four hours. A good tooth-powder can occasionally be used with benefit. A toothpick, made of wood, ivory, or quill, or of some other substance softer than the tooth itself, should be used after each meal. A hard toothpick, like a pin or needle, is likely to scratch the enamel of the tooth, and thus cause it to decay.

Why is it a bad practice to crack nuts with the teeth, or to bite other hard substances?

The cracking of nuts, or even the constant biting of thread, or any hard substance, is injurious to the teeth because it is liable to break the hard enamel of the teeth and thus expose the inner and softer portions to the action of the air and the liquids of the mouth, causing decay.

Why should decaying teeth be filled or removed?

If the decay of a tooth is not stopped, it will soon begin to cause pain, and its service in preparing food will finally be lost. If taken in season and properly filled, its usefulness may often be prolonged many years. Besides this, a decaying tooth is a filthy thing to have in the mouth. Its presence there taints the breath, and produces decay in others; consequently, if a tooth is past filling, it should be removed at once.

NOTES AND SUGGESTIONS.

Note 1. Power of Bones to Resist Decay.—Bones and teeth have a remarkable power of resisting decay. They remain long after all other parts of the body have wasted away. Bones have been found which belonged to animals that lived before man appeared on the earth. The teeth resist decay even longer than the bones.

2. *Weight of Skeleton.*—The weight of the skeleton is about one-tenth that of the entire body; consequently, the weight of the bones of a man of 140 lbs. would be about 14 lbs.

3. *Strength of Bones.*—The strength of human bones when used as levers, as compared with other substances, is remarkable. Bones containing the same amount of material are twice as strong as oak timber.

4. *Chemical Composition of Bone.*—Of 100 parts of bone, 33 are

organic substance or cartilage, 57 are phosphate of lime, 8 are carbonate of lime, and the other 2 are fluorid of calcium and phosphate of magnesia.

5. *Proof of Animal Matter.*—Immerse a bone for a few days in dilute muriatic or sulphuric acid. The acid will eat out all the lime, or earthy matter, leaving the cartilage, which will be so soft and flexible that it may be tied in a knot. The experiment can be easily tried with the bone of a chicken's leg.

6. *Proof of Mineral Matter.*—Place a bone for a short time in a hot fire. All the animal matter will be burned out, leaving the lime, or mineral matter. This will be white and brittle, and easily crushed to a powder.

7. *Proof that the Blood Circulates through the Bones.*—If madder be mixed with the food of an animal, the bones soon become red.

8. *Suggestions.*—The fresh bones of the leg of a sheep or calf can readily be obtained from the market. Everything that can be learned objectively by examining the specimens should be learned before assigning a lesson from the text-book. The internal structure can be examined by sawing the bone longitudinally and cross-wise. Have also before the class a dry bone, and let pupils note the differences between this and the fresh one. If possible, examine the structure of bone with a microscope. Glass slides containing specimens already prepared can be had for a few cents. If the school is not provided with a skeleton, the structure of the skull, joints, etc., can be understood by examining the skull of a rabbit or cat. It will be of interest to compare the human skeleton with that of other animals. Some animals have the skeleton outside; as, crabs, lobsters, oysters, clams, etc. The skeleton of the turtle is peculiar. The teeth of various animals may be compared. In doing so, note the relations between the form and structure of the teeth and the kind of food upon which the animal lives.

CHAPTER VII.

FOOD AND DRINK.

What is food?

A food is any substance, solid or liquid, which, when taken into the body, repairs the waste or prevents the loss of any of its tissues.

Why do we need food?

We need food to repair the waste caused by the wearing out of tissues, to keep up the warmth of the body, and in growth to supply material for new tissues.

What are the evidences that the tissues of the body waste, or wear out?

If little or no food be taken for a time, the body, as we say, becomes thin and poor, and gradually loses in weight. If we are compelled to continue to do without food, the body continues to lose weight, until finally death ensues. Loss in weight simply means loss, or waste, of tissues.

What are the causes of waste?

Motion everywhere is followed by waste. The stream wears away its banks; the steam-engine uses up coal and water to produce motion, and is itself worn out by the motion produced. The burning of coal in producing heat is called *chemical action;* the wear caused by the motion of the machinery is due mostly to *friction.* So it is with our bodies: the motive-power is furnished by the food we eat; but, unlike the engine, our food is not burned in a separate part of the body. The food becomes a part of the body, and, therefore, it is the whole body that is slowly burning. Nevertheless we can say, as of the engine, that the causes of waste are *chemical action* and *friction*.

How do we know that food keeps the body warm?

If we go out on a cold day without having eaten a sufficient amount of food, we feel chilly and cold, when with a hearty meal we would be comfortable and warm. We need more and heartier food in winter than in summer. Thus we all know that mother's advice on a cold day is correct, when she says, "Eat a hearty breakfast before you go out, or you will be cold."

Why does a child need more tissue-making food than a grown person?

A grown person needs only the food necessary to renew worn-out tissues, while a child must have not only food enough to renew the worn-out tissues, but enough more to make the new tissues required for his growth.

Why do we need more than one kind of food?

In order to keep the body warm, and to provide for the repair and growth of all the tissues of the body, the food must contain the proper amount of heat-producing material, as well as the materials which make bone, muscle, fat, etc. All these in the right proportions are seldom found in any one article of food.

Into what four divisions, or classes, may articles of food be divided?

For convenience, articles of food may be divided into four classes: 1st, *nitrogenous foods*, or *albumens;* 2d, *starches* and *sugars;* 3d, *oils* and *fats;* 4th, *inorganic*, or *mineral foods*.

Of what use are nitrogenous foods, or albumens?

The nitrogenous foods, or albumens, are especially muscle-making foods. They are called nitrogenous foods because they are rich in nitrogen, an essential element in lean meat. They are called albumens, or albuminoids, because they contain a white substance called albumen. The white of an egg is pure albumen. The foods belonging to this class contain to some extent the elements necessary to form other kinds of tissues, yet we say that their particular use is to make muscle.

Name the principal articles of food which may be classed as muscle-making foods, or albuminoids.

Among the albuminoids, or the articles of food most useful for making muscle, are eggs, lean meat, milk (the cheesy part, or curd), peas, beans, wheat, barley, oats, and other grains.

Of what use are starches and sugars?

The starches and sugars contain carbon, oxygen, and hydrogen, but no nitrogen. Their use is to serve as a kind of fuel which is consumed by the oxygen in the body. The heat thus produced keeps the body warm, and gives to it the energy, or strength, by which it is able to perform its work.

Name the articles of food which are rich in starch or sugar.

Starch is the important food substance in all vegetables. It is found abundantly in potatoes, rice, corn, wheat, barley, tapioca, arrow-root, etc. The sugars are derived from vegetables. They are used for food in a nearly pure form, as sugar, candy, honey, etc., and are also taken to a small extent with milk and vegetables such as melons, beets, peaches, strawberries, grapes, pineapples, etc.

Of what use are fats and oils?

The use of the fats and oils is nearly the same as that of the starches and sugars; that is, to produce heat and vital energy. Fats and oils produce about twice the amount of heat, or energy, as that produced by an equal amount of sugar or starch.

From what foods are fats and oils obtained?

Fats and oils are obtained mostly from fat meats, eggs, butter, and milk. They are also obtained in small quantities from nuts, olives, beans, and other vegetable foods.

What are some of the inorganic, or mineral, substances found in the body?

Among the inorganic, or mineral, substances found in the body are water, lime, salt, iron, phosphorus, potash, and sulphur.

Of what use is water as a food?

Water is present in all the tissues and in all the fluids of the body. About seventy per cent. of the weight of the body is water. Water keeps the blood thin so that it will flow more readily; it helps to carry off waste matter; it aids in digesting food; and keeps the bones, muscles, and other tissues from becoming hard and dry. The body requires on an average about four and a half pounds of water per day.

Of what use are other inorganic, or mineral foods?

Lime is required by the body for bones and teeth. Salt is found in all the tissues of the body except the enamel of the teeth; it is useful in aiding the digestion of other substances. Iron and potash are found in the blood; phosphorus and sulphur are found in the muscles and other tissues.

From what sources are inorganic, or mineral, foods obtained?

Water, besides being taken into the body in its pure form, is taken in with every kind of food, whether solid or liquid. For instance, bread is 37 per cent. water; fat beef, 51 per cent.; lean mutton, 72 per cent.; potatoes, 75 per cent.; turnips, 91 per cent. Salt is used as a seasoning for other foods. Lime is found in hard water, also in meat, milk, and vegetables. Iron is found in milk and eggs, and the same may be said of phosphorus and sulphur. Potash exists in vegetable matter.

Why does the quantity of food which a person requires vary?

From what we have already learned we may rightly conclude that a man requires more food in youth than in old age, more during an active than a sedentary period of life, more in a cold climate than in a warm one, more in winter than in summer.

What is the estimated daily amount of food and drink required by a healthy man of active habits?

The amount of food required daily by a healthy man of active habits is estimated to be about six pounds. Supposing his food to consist of beefsteak, bread, potatoes, butter, and water, leaving out tea, coffee, and salt, the proper quantity of each would be about as follows: beefsteak, 8 oz.; bread, 20 oz.; potatoes, 30 oz.; butter, 1 oz.; water, 37 oz.

CHAPTER VIII.

ORGANS OF DIGESTION.

Name the organs of digestion.
The organs of digestion are: the mouth and salivary glands, the stomach, the pancreas, the liver, and the intestines.

What is the alimentary canal?
The alimentary canal is the canal, or tube, formed by the organs of digestion, together with the passages which connect them.

Describe the alimentary canal.
The alimentary canal is about twenty-five feet in length. Its walls are composed mostly of muscles, and it extends from the mouth downward through the body. It is lined throughout its entire length by a thin, soft, moist, reddish skin, or membrane, called the *mucous membrane.* In and around it are hollow organs, called glands, which pour into it certain fluids, which help to dissolve the food and to fit it for entering the blood.

Through what four processes must the food pass in order to build up, or nourish, the body?
In order to give nourishment to the body the food must pass through four processes. First, it must be dissolved, or changed into a liquid in the alimentary canal, called *digestion;* secondly, it must pass from the alimentary canal into the blood-vessels, called *absorption;* thirdly, it must

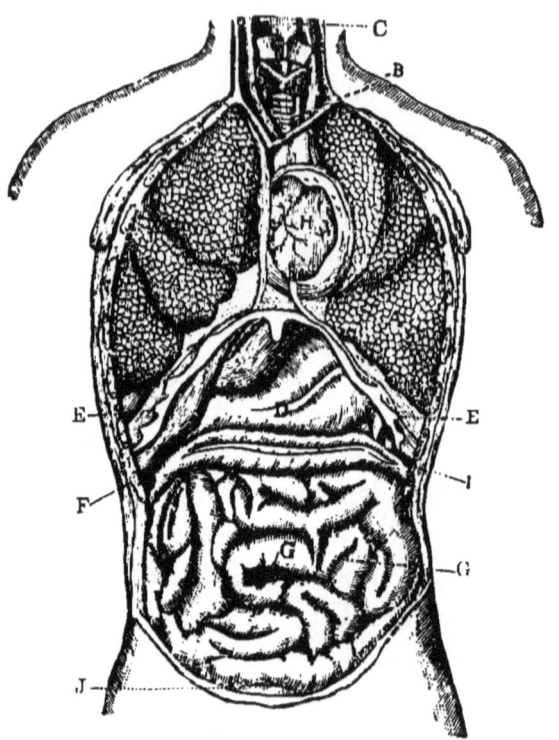

INTERNAL ORGANS.

B, the *trachea* (windpipe).
C, *œsophagus* (gullet).
E, *diaphragm*.
F, *liver*.
I, *spleen*.
D, *stomach*.

G, *intestines*.
H, *heart*, the pericardium being laid open.
A, *lungs*.
J, *bladder*.

be carried to all parts of the body by the blood, called *circulation*; fourthly, it must be taken from the blood and changed into tissues, bone, muscle, fat, etc., called *assimilation*.

What is the object of digestion?

The object of digestion is to change solid food into a liquid, and to prepare all food for entering the blood. It is plain that food, in the condition in which it is eaten, could not be taken into the blood. There is no open passage-way from the alimentary canal to the blood-vessels. No food can get out of it unless it soaks through its walls, and in order to do this, it must be changed into a liquid.

Of what use are the muscles in the walls of the alimentary canal?

The muscles in the walls of the alimentary canal aid digestion in two ways. They force the food along, and knead it and mix it with the juices.

What part of the work of digestion is performed in the mouth?

In the mouth the food is cut and ground by the teeth to such fineness that it can be readily swallowed and easily acted upon by the juices of the stomach. It is also mixed with the saliva of the mouth, which aids in swallowing, brings out the taste, and is a help to digestion.

What is the office of the salivary glands, and where are they situated?

The office of the salivary glands is to secrete, or take from the blood, the fluid called *saliva*, which moistens the mouth. When we eat or taste something which we like, so much saliva is sometimes poured out that we say the "mouth waters." There are three pairs of salivary glands. The largest are called the *parotid glands*. They are situated one in front and below the lower part of each ear. They are the glands which swell when we have the mumps. One of the other pairs is situated under the tongue; they are called the *sublingual glands*. The other pair is situated under the lower jaw; they are called the *sub-maxillary glands*. From each of these glands there are little tubes, called *ducts*, leading into the mouth.

How does the saliva aid digestion?

The saliva moistens and softens the food, and to some extent dissolves it. It also acts upon the starchy matter in the food by changing it into a kind of sugar, which is quickly dissolved on entering the stomach.

Describe the œsophagus.

The *œsophagus* (œ-soph'-a-gus) is a muscular tube about nine inches long, leading from the mouth to the stomach. The muscles in the walls of the œsophagus extend round and round in a ring-like form, so that when one ring contracts it crowds the food on to the next. In this way it is pushed on until it reaches the stomach.

What is the shape and size of the stomach?

The *stomach* is a pear-shaped bag, or pouch, which will hold about two quarts. When empty, it collapses, like any other bag. When full it is some ten or twelve inches long, and about four inches broad.

Where is the stomach situated, and how many openings has it?

The stomach is situated across the upper part of the abdomen, directly under the diaphragm, the larger end to the left side. It has two openings. The opening where the food enters is called the *cardiac* orifice. The other opening at the right end, where the food leaves the stomach and where the intestines begin, is called the *pylorus* (py-lo'-rus),

ORGANS OF DIGESTION. 41

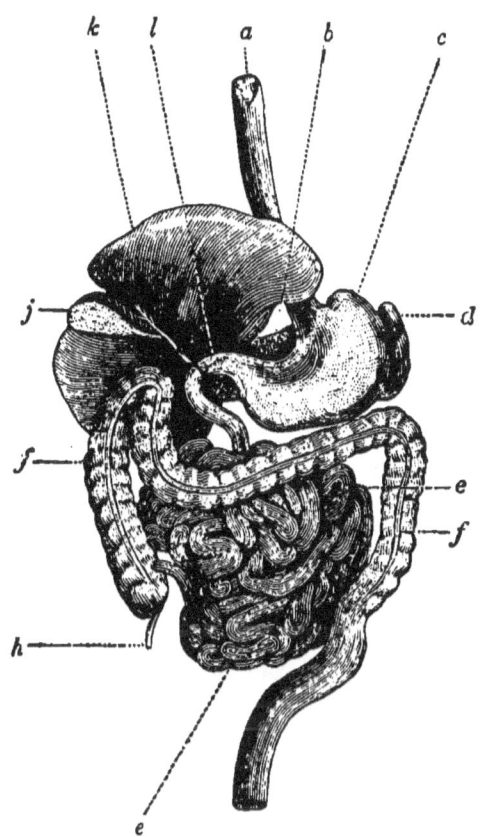

THE ORGANS OF DIGESTION.

a, the œsophagus.
b, the pancreas.
c, the stomach.
d, the spleen.
e, the small intestine.

f, the large intestine.
h, the appendix.
j, the gall-bladder.
k, the liver.
l, the pylorus.

or gate-keeper. It is so called because it closes so as to keep the food in the stomach until it is ready to pass out.

Describe the walls of the stomach.

The walls of the stomach are thin, soft, and flexible. They are composed of three coats, or layers. The outer coat is a layer of tough, fibrous tissue; its use is to give strength and firmness, and to hold the stomach in place by being attached to the back and sides of the abdomen. The middle coat is a *muscular* coat; its use is to give motion to the contents of the stomach, keeping them well mixed. The inner coat of the stomach is called the *mucous* coat. It contains many blood-vessels, and the glands which constantly supply the stomach during the process of digestion with an important fluid called *gastric juice*.

Describe the gastric juice.

The gastric juice is a clear, almost colorless, fluid, with a sharp acid taste. It contains a peculiar substance, called *pepsin*. The acid and the pepsin are both necessary to the digestion of food in the stomach. The quantity of gastric juice which daily finds its way into the stomach varies considerably. It is estimated to be between five and fourteen pounds.

CHAPTER IX.

ORGANS OF DIGESTION.—CONTINUED.

What is the office of the gastric juice?
The principal office of the gastric juice is to dissolve the class of foods which we have called albuminoids; such as lean meat, cheese, eggs, and, in general, all animal matter. It has comparatively little effect upon starchy matter.

What is the food in the stomach called after it has been acted upon by the gastric juice?
The food in the stomach, after it has been acted upon by the gastric juice, is called *chyme*. Chyme is a pulpy, soup-like substance, of a grayish color. In it are fats and oils, as well as starchy matter, which have not yet been made soluble.

How does the digested food escape from the stomach?
When the food has been properly changed into chyme, the muscles at the pylorus relax, and the partially digested food passes out in little jets into the *duodenum*, which is the first portion of the small intestines. The duodenum is so called because its length is about equal to the breadth of twelve fingers.

What fluids are mixed with chyme after it enters the duodenum?
When the chyme enters the duodenum, two juices, the *bile from liver* and the *pancreatic juice*, are poured in and mixed with it. The walls of the intestines also secrete a

digestive fluid, which is mixed with the chyme as it passes along.

What seems to be the use of the bile?

The chief use of the bile seems to be to digest the fats and oils in the food, upon which the gastric juice does not act. It also helps to separate the nutritious from the useless parts of the food.

What is the use of the pancreatic juice?

The use of the pancreatic juice seems to be very much like that of the saliva. It acts chiefly upon starchy matter which has escaped the saliva, and changes it into sugar. It also helps to dissolve albuminoids which have not been digested by the gastric juice.

Of what use is the intestinal juice?

The *intestinal juice* is the fluid which completes the work of digestion. It seems to combine in one the properties of all the other digestive fluids. Like the saliva, it acts upon starchy matter; like the gastric juice, it is a solvent for albuminoids; and like bile, it has the power of rendering soluble fats and oils. Thus it seems that nature intends that this last fluid in the digestive tube shall perform all the work left undone by the others, and so prevent the waste of any part of the food.

After the food in the intestines has been acted upon by all the digestive fluids, what is it called?

After the food has been acted upon by all the digestive fluids, it is called *chyle*. Chyle is a white, milky liquid, and it contains in solution all the digestible portions of the food.

Through what process must the food next pass?

The digested or liquid food, called chyle, must now pass out of the alimentary canal into the blood. As has already been said, this process is called *absorption*.

ORGANS OF DIGESTION.—CONTINUED.

How does the digested food pass out of the alimentary canal into the blood?

Digested food passes from the alimentary canal into the blood in two ways. First, *by the minute blood-vessels in the walls of the stomach and intestines.* Secondly, *by the lacteals,* which are small tubes leading from the inside of the intestines.

How does the absorption of food take place by blood-vessels?

The walls of the digestive canal, especially of the small intestines, are full of blood-vessels, and the liquid food rapidly soaks through them and enters immediately into the veins and capillaries. These capillaries and veins unite and form a large vein called the *portal vein,* which carries the blood containing digested food to the liver. After passing through the liver, it is gathered up by other veins, and so finds its way into the right side of the heart.

How does absorption of food take place by the lacteals?

Opening through the inner lining of the intestines are myriads of little tubes called lacteals. These tubes are so called from a Latin word meaning "milky," and because they carry a white, milky fluid. These little tubes take the chyle from the intestines. After leaving the intestines they unite into larger tubes, until finally they form a single tube about the size of a goose-quill, called the *thoracic duct.* The thoracic duct carries the fluid upwards, along the back-bone, and pours it into a large vein in the neck, which empties it into the heart.

After the digested food enters the blood, how is it carried to the different parts of the body?

After the digested food enters the blood it becomes a part of the blood, and is carried in the circulation to all parts of the body, where the different elements in it are left wherever they are needed for the repairing or building up of tissues. This completes the third process through which the food must pass in giving nourishment to the body.

What takes place in the fourth process, called assimilation?

In the process of assimilation each cell and tissue of the body takes from the blood and adds to itself the material necessary to its own nourishment and growth. Thus food is converted into bone, muscle, fat, skin, etc.

What time is required to properly digest food?

The time required to properly digest food varies from one to five hours. The health of the individual, and the amount, kind, and quality of food, are some of the causes which make a difference in the time required for digestion.

Name some kinds of food which are quickly digested.

Among articles of food which are quickly digested are boiled rice, barley soup, sweet apples, raw eggs, tapioca, milk, custard, and oysters.

Name some kinds of food which are not quickly digested.

Among articles of food which are not quickly digested are fat pork, beef suet, boiled cabbage, cheese, fried beef, etc., and any others which are so compact that the digestive fluids will not readily enter and dissolve them.

CHAPTER X.

HYGIENE OF ORGANS OF DIGESTION.

Mention some things which must be observed in order to keep the organs of digestion in a healthy condition.

In order to keep the organs of digestion in a healthy condition, we should be careful always to take as nearly as possible the right quantity of food, not too much or too little; it should be good in quality, not indigestible, stale, or adulterated; it should be of the proper temperature, not too hot or too cold; it should be properly chewed and mixed with the saliva of the mouth; the juices of the mouth and stomach should not be too much diluted by drinking during meals; the juices of the mouth and stomach should not be wasted or destroyed by chewing gum or tobacco, or by the use of alcohol, or food too highly seasoned; food should be taken at proper intervals; pure air and exercise are also necessary to a keen appetite and good digestion.

What harm is likely to come from not eating enough food?

Every action causes some of the tissues of the body to waste away, or wear out. If the amount of food taken is not sufficient to repair the waste, the body loses weight, grows thin, becomes feeble, and finally diseased.

Why is the quantity of food which a person requires greater during the period of growth?

The quantity of food which a person requires is greater during the period of growth, because the food must not only repair the waste, but supply material for building new tissues. The same thing is true when the body has been emaciated by starvation or disease.

How does exercise affect the quantity of food required?

When exercise is increased or diminished, the quantity of food should be accordingly increased or diminished. This is true because exercise causes waste, and the quantity of food needed to repair the waste will depend upon the amount of exercise.

Why is more food required in winter than in summer?

A considerable portion of the food which we eat is expended in keeping the body warm. The body has nearly the same temperature summer and winter; but it would become colder in winter, if we did not eat more, or in other words burn more fuel in it, for the same reason that more fuel is required to keep our rooms warm in winter than in summer. Cold climates, likewise, demand more heat-producing food than warm ones. The Esquimau is fond of fats and grease, while the native of the tropics lives almost wholly on vegetables.

Why should care be exercised in regard to the quantity of food during sickness?

Sickness, or disease, means that some organs of the body are not able to do their usual amount of work. The failure of one set of organs affects, more or less, all others. Consequently sick persons have poor appetites, and great care has to be taken that the organs of digestion are not overtaxed by too much food. If the digestive organs are not able to supply the necessary fluids, and to perform the whole work of digestion, the food sours, and being imperfectly digested irritates the lining of the intestines, and thus weakens the body instead of strengthening it.

What is the harm of eating too much food?

The glands of the stomach are only able to secrete gastric juice enough to dissolve a certain amount of food. If more than this amount is eaten, a part of it, at least, must go imperfectly digested. Besides this, too much food gives the organs of digestion too much work to do, and as a natural consequence the glands are weakened by excessive action.

Why are immature or over-ripe vegetables not good articles of food?

Immature fruits and vegetables are not usually good articles of food, because they are likely to be indigestible, and they lack the fully formed juices which make vegetable foods most valuable. In over-ripe fruit and vegetables the juices have decomposed and have formed other substances which are harmful.

Why is it not safe to eat raw pork?

It is not safe to eat ham, sausages, or any other form of pork in a raw or imperfectly cooked condition. The muscle of the pig often contains a minute animal parasite, or worm, called *trichina spiralis*. If taken alive into the human body, this worm multiplies with great rapidity and causes a painful and serious disease. The life of the parasite and its ability to do serious harm are destroyed by proper cooking.

What is the object of cooking food, and why should food be well cooked?

The object of cooking food is to improve its flavor, to make its temperature more nearly equal to that of the body, to soften it so that it can be more easily broken up by the teeth and more easily acted upon by the digestive fluids. The proper cooking of food is a matter of great importance.

By improper cooking, its flavor may be spoiled, it may be made tough and indigestible, or its nutritive qualities may be destroyed. The best food may be made unhealthful by improper cooking.

Why should food be eaten at a proper temperature?

If food or drink be taken too hot, it is liable to injure the mucous membrane of the gums, mouth, and stomach. Such practice is a fruitful cause of soft gums, decayed teeth, sore mouth and indigestion. If food or drink is taken too cold, the cold food or drink takes so much heat from the stomach that the process of digestion is delayed. The natural temperature is about 100° F. To drink freely of ice-water might lower the temperature to 40°. It would take, perhaps, half an hour for the stomach to regain its natural temperature. In the mean time the process of digestion would go on very slowly, and the food might become sour.

Why should food be properly masticated?

Food should be properly chewed, or masticated, in order that it may be well mixed with the saliva, and that it may be in a condition fine enough to be easily acted upon by the gastric juice and other fluids. Rapid eating is a violation of good manners, and of the laws of hygiene.

Why is it harmful to take much drink with food?

The natural fluid with which to moisten food in the mouth is the saliva. The saliva is an important digestive fluid, and, if any other fluid, as water, tea, or coffee, be used as a substitute, digestion will not be so rapid or perfect. Besides, the water will so dilute the gastric juice that the work of digestion will have to stop until a portion of the water has been absorbed.

What harm to digestion comes from chewing gum or tobacco?

The chewing of gum, tobacco, or any other substance

excites a flow of saliva into the mouth. This saliva, not being used for moistening or digesting food, is wasted; and besides, the glands, not being able to act constantly, are unable to supply more when needed. Indigestion is liable to be the result.

What are some of the effects of alcohol upon the organs of digestion?

Many cases of dyspepsia are due to alcoholic drinks. Such drinks irritate the lining of the stomach and stimulate the flow of the gastric juice; but the alcohol absorbs the water from the gastric juice, and so acts upon it as to diminish its solvent power, thus seriously interfering with the process of digestion. If the use of alcohol is great and long continued, a chronic inflammation of the inner coating of the stomach occurs; the walls become thick and hard, and traces of ulceration are often found. The liver is also very liable to disease from the use of alcoholic drinks.

Why should condiments and spices be used sparingly?

Condiments are substances which are used to sharpen the appetite, to give relish for food, and to stimulate the digestive organs. Pepper, mustard, horse-radish, pickles, fancy meat-dressings, and highly-seasoned sauces, may be classed as condiments. Their immoderate use injures the taste, inflames the mucous membrane, excites an excessive secretion of the digestive fluids, and causes the consumption of more food than the body requires.

Why should food be taken at regular intervals?

The organs of digestion, like all other organs of the body, must have periods of rest. Like all other organs, too, they do their work best when their tasks are done at regular periods. Eating between regular meals is a habit ruinous to the digestive organs, because it disturbs the chemical processes which take place, and gives the stomach no time to rest. It is a habit, in many respects, similar to that of over-eating.

Why is it well to rest for a little time just before and after eating?

When any part of the body is being vigorously exercised, the flow of blood and the nervous forces of the body are especially directed towards that part. This concentration cannot be suddenly changed to some other set of organs, consequently a little time should be given just before meals for gradually restoring the system to its natural condition. For the same reason, severe exercise and hard study just after a full meal are very apt to hinder digestion. After a full meal the vital forces of the body are engaged in helping the stomach digest food. If, in addition to this, they are required to help the muscles or brain, one or both will get imperfect service. Moderate exercise of the muscles, pleasant conversation, and a hearty laugh, are aids to digestion.

Why does good digestion require pure air?

The organs of digestion in order to do their work well must have a plentiful supply of pure blood. In order to have pure blood we must breathe pure air. Hence poor ventilation is a frequent cause of indigestion. People who sleep in ill-ventilated rooms have little or no appetite for breakfast.

NOTES AND SUGGESTIONS.

Note 1. Collection of Specimens.—The principles which underlie the matter of our daily food cannot be too well understood. Children should be made familiar with all classes of food-stuffs, and the valuable elements in each. An interesting and valuable collection of specimens of such articles can be easily made. The specimens can be kept in glass bottles and should be neatly labelled. In class let pupils handle them, taste them, talk about them.

NOTES AND SUGGESTIONS. 53

2. *Show the Organic Substances in the Albuminoids.*—The subject of foods and digestion, also, admits of many useful and interesting experiments. The albuminoids, or nitrogenous foods, contain one or more of the following organic substances : albumen, caseine, fibrine, gelatine, gluten, and legumen. Albumen may be shown as the white of an egg ; also by washing and squeezing a piece of lean meat in cold water with a lemon-squeezer, until nothing but a whitish stringy mass is left. The albumen has been dissolved in the water and may be coagulated by heating the water. The whitish stringy mass which was left is *fibrine.* To show caseine, pour some dilute acid, vinegar, or liquid rennet, into some milk. A whitish curd will be formed. This is *caseine,* the chief constituent of cheese. Gelatine may be obtained by boiling a bone for a long time. For instance, calves' feet are boiled to make jelly for the sick-room. Glue is simply gelatine obtained by boiling the bones, hoofs, horns, etc., of animals. Gluten may be shown by putting a small quantity of flour in a cloth and squeezing it in a basin of water. The *gluten* is the sticky substance remaining in the bag. If the water is allowed to stand, the *starch* will settle to the bottom as a white powder. If beans or peas in the pod be boiled for a time, they will finally become a sticky, pulpy mass. This pulpy mass is *legumen.* It resembles albumen in the egg, and gluten in flour.

3. *Show Starch in Vegetable Foods.*—The test for starch is a little tincture of iodine. This dropped into any mixture containing starch will turn it blue. Boil a little flour, rice, tapioca, or potato, in a test-tube, and add a drop of the tincture of iodine. It will quickly turn blue. In regard to specimens of sugars and oils, no suggestion is necessary. They can easily be obtained.

4. *Prove that Milk contains All the Necessary Elements for Food.*—A calf fed upon milk alone will grow and thrive. He obtains from it all the materials necessary for making bones, muscles, fat, hair, skin, etc. That milk is a compound food may be shown by experiment. Take some milk "fresh from the cow," place it in a tall glass vessel and let it stand for a few hours. *Cream* will then be found on the top. Take off the cream, and to the milk which is left add a little liquid rennet, acid, or vinegar, and a whitish *curd* will be found, leaving a watery liquid called *whey.* The milk now has been separated into three parts, cream, curd, and whey. Cream is the *fat* in milk ; curd is the caseine, or *nitrogenous part;* and whey is *milk-sugar* and *mineral matter* dissolved in water.

5. *Saliva changes Starch to Sugar.*—Place in the mouth a small quantity of starch. Chew it slowly until it becomes thoroughly moistened. Notice that it has a sweet taste. The same thing can be shown with a cracker or a piece of bread.

6. *The Mouth waters at the Thought or Sight of Food.*—Take in your hand or think of some favorite article of food, especially if you are hungry, and note the flow of saliva. This is forcibly illustrated by cows, when they sometimes stand on the opposite side of a fence waiting for food which is being prepared in sight of them. The flow of saliva is often so great that it almost pours from their mouths in streams.

7. *Helps for showing the Anatomy of Digestion.*—The position and arrangement of the organs of digestion can be quite well understood by the use of a manikin, or charts, or by a diagram on the blackboard. A good idea of the structure of the walls of the stomach may be obtained by examining a piece of a pig's stomach, which somewhat resembles the human stomach. Such a specimen can easily be obtained from the butcher. The liver and pancreas can be obtained and examined in the same way.

8. *How a Knowledge of Many Facts regarding Digestion was obtained.*—In 1822 a French Canadian, Alexis St. Martin, was wounded by a gun-shot which tore away the flesh of the abdomen and made a hole in the stomach. St. Martin recovered, but the hole did not close. A piece of flesh hung over the opening which could be lifted up so as to permit a look into the stomach. The physician, Dr. Beaumont, tried many experiments upon St. Martin, and by these experiments many things relating to digestion were found out. Experiments upon animals have also added greatly to our knowledge of the laws of digestion.

9. *Comparative Anatomy of Organs of Digestion.*—The teacher can orally supplement the lesson with many interesting facts by comparing the digestive organs of various animals. The peculiarities of the digestive apparatus of the cow, the hen, the snake, will especially be of interest.

10. *Effect of Alcohol on Albuminous Food.*—Put the white of an egg into a small quantity of alcohol. Stir the egg and the alcohol together for a few minutes. The egg will begin to harden and have a cooked appearance. The effect of alcohol upon such food is to hinder, not to help, digestion.

CHAPTER XI.

THE BLOOD AND ITS CIRCULATION.

What is the blood?

The blood is the red liquid which circulates through the different parts of the body. Blood has a saltish taste. It is a little heavier and thicker than water.

What are the uses of the blood?

The uses of the blood are, to carry digested food to all parts of the body, to keep up the warmth of the body and to provide it with moisture, to gather and convey waste matter to places where it may be discharged, and to carry oxygen to such tissues as require it.

Does the blood circulate through every part of the body?
What is the average quantity of blood in the human body?

The blood circulates through every part of the body except the outer part of the skin, the nails, enamel of the teeth, and cornea of the eye. The average quantity of blood in the human body is about eighteen pounds.

Of what is the blood composed?

The blood is composed of an almost colorless fluid called *plasma*, in which float a great multitude of little circular bodies, or discs, called *corpuscles*. The corpuscles are of two kinds, *red* and *white*. There are about four hundred times as many red corpuscles as white ones. The individual corpuscles can be seen only by the aid of a microscope. The red corpuscles are so numerous and so closely packed together that they make the entire substance of the blood to appear to be red. It is said that five million corpuscles will float around in a single drop of blood.

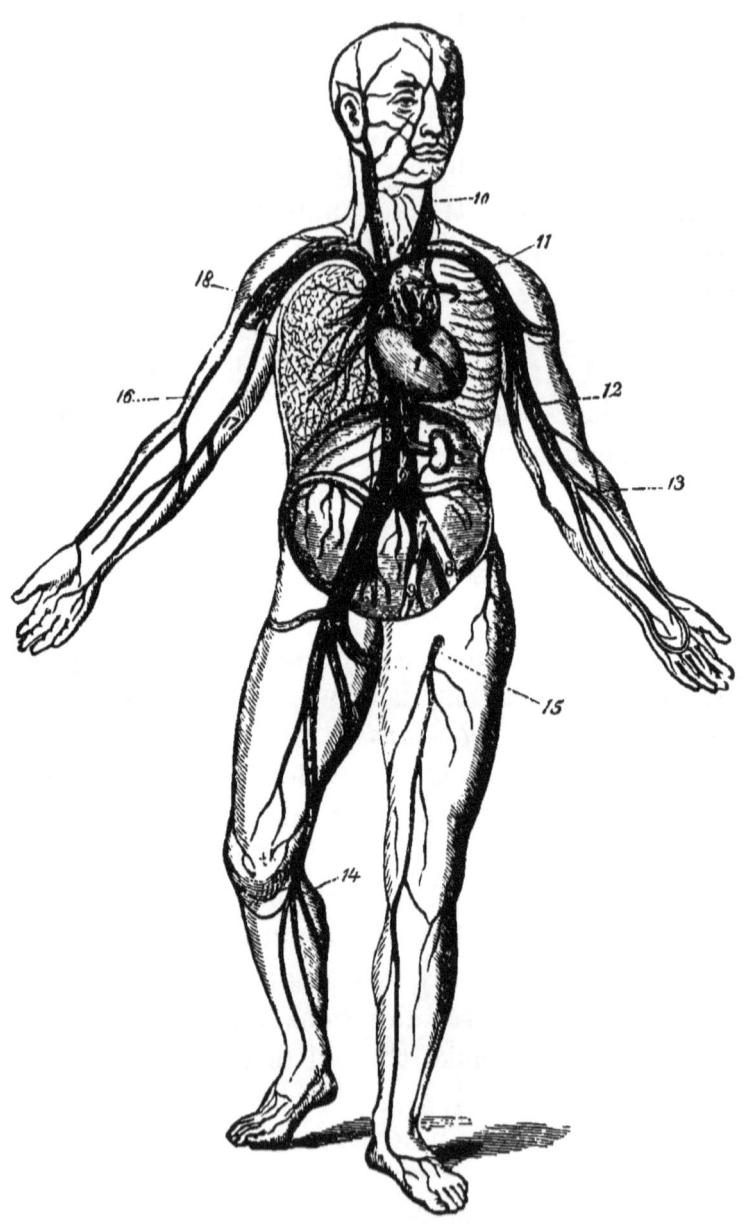

VIEW OF THE HEART WITH THE CIRCULATORY APPARATUS.

1, the heart ; 2, the pulmonary artery ; 3, the inferior or ascending vena cava ; 4, the superior or descending vena cava ; 5, the aorta ; 6, its point of division ; 7, point of division of left common iliac artery ; 8, external, 9, internal, iliac arteries ; 10, left common carotid, with internal jugular vein lying to its outer side ; 11, left sub-clavian artery and its continuation, the axillary ; 12, the brachial, 13, the radial, 14, the anterior, tibial arteries ; 15, long saphenous vein ; 16, cephalic vein of right arm ; 17, basilic vein ; 18, ramifications of pulmonary vein and arteries in the left lung. The direction of the blood current is indicated by the arrows. The veins are darker than the arteries.

Of what is the plasma composed?

The plasma, or nutritive fluid, is composed of water richly laden with materials derived from the food. It, in fact, contains all the elements of which the body is composed.

Of what use are the red corpuscles?

The red corpuscles are carriers of oxygen. They take oxygen from the air in the lungs and carry it for distribution to the various tissues of the body. The office of the white corpuscles is not certainly known. They are supposed to change after a little into red ones.

What takes place when blood is exposed to the air? Of what advantage is this?

If blood is exposed to the air, it immediately clots, or *coagulates.* That is, it thickens and assumes a semi-solid form. This coagulation of the blood saves us from bleeding to death when we are wounded. The clots which form stop up the mouths of the cut blood-vessels.

What is the effect from loss of blood?

Loss of blood causes weakness, and, if very great, fainting. During a fainting-fit the heart nearly stops its action, which makes the blood flow more slowly, and gives it a chance to coagulate. Thus a fainting-fit may save life. A man's body contains six or eight quarts of blood. The loss of more than half of this would surely cause death, and a very much smaller amount might do so.

What is meant by the circulation of the blood?

The blood is in constant motion during life. It is continually flowing from the heart toward the various organs of the body, and from these organs back to the heart again. This movement is called the *circulation of the blood.*

By whom was the circulation of the blood discovered?

The circulation of the blood was discovered by Dr. William Harvey, an Englishman, in the year 1619.

What length of time is required for the complete circulation of the blood?

In some cases a drop of blood goes the complete round of the body in about twenty-two seconds. The entire blood in the body probably completes this circuit inside of two minutes.

CHAPTER XII.

THE ORGANS OF CIRCULATION.

Name the organs of circulation.

The organs of circulation are the *heart*, the *arteries*, the *capillaries*, and the *veins*. As we study these organs, we shall find that the heart is a central forcing-pump, and that the arteries, veins, and capillaries are pipes which either carry the blood from the heart or back to it again.

Describe the heart.

The heart is a hollow, muscular organ, pear-shaped, and situated in the centre of the chest, above the diaphragm, with the small end pointing down, and to the left. It is about as large as the closed fist. It is a double organ, having a partition-wall extending through the centre from top to bottom, which separates the right side from the left. Each of these sides has two hollow chambers or cavities. The upper cavity on the right side is called the *right auricle;* the upper one on the left side is called the *left auricle.* The lower cavities, in the same way, are called the *right ventricle* and the *left ventricle.* The walls of the ventricles are thicker and stronger than the walls of the auricles, and those of the left ventricle are much stronger than those of the right ventricle. The right auricle opens by a valve into the right ventricle, and the left auricle opens by a valve into the left ventricle; but there is no connection between one side of the heart and the other. The heart is, in fact, two pumps, each with its own work to do.

Describe the valves of the heart.

The openings between the auricles and ventricles are provided with little swing-doors, called *valves*, which open to allow the blood to flow into the ventricles; but close, if it attempts to flow back into the auricles. The valve between the right auricle and the right ventricle is composed of three thin flaps of tough membrane, and is called the *tricuspid* (three-pointed) valve. The valve between the left auricle and the left ventricle consists of two such flaps, and is called the *mitral* valve, from the supposed similarity to a bishop's mitre. The passages from the ventricles into the arteries are closed by half-moon-shaped valves, called *semi-lunar* valves.

By what is the heart surrounded?

The heart is surrounded by a loose sac or membrane, called the *pericardium*. The pericardium is soft and smooth, and gives out an oily fluid which prevents friction and keeps it moist and pliable.

What blood-vessels are connected with the heart?

Two large veins pour the impure blood as it returns from all parts of the body into the right auricle. From the right ventricle the *pulmonary artery* carries this blood to the lungs to be purified. Into the left auricle the *pulmonary veins* pour the pure blood as it returns from the lungs. From the left ventricle a large artery, called the *aorta*, takes the blood, and, through its various branches, distributes it to all parts of the body.

Describe the arteries.

The arteries are tube-like canals which carry blood away from the heart. Their walls are made of tough fibrous materials, so that they can endure the necessary strain without being broken. The walls also are elastic, and thus they aid the heart in forcing the blood along. One artery (the pulmonary) with its branches conveys the blood to the lungs. The other (the aorta) with its branches extends to every part of the body. These branches as they get farther from the heart become smaller and more numerous, until finally they form throughout the entire body a complete network of hair-like tubes, called *capillaries*. They are, however, smaller than any hair, and can only be seen by the use of a microscope.

Describe the veins.

The veins are the return-pipes which take the blood back to the heart. They start from the capillaries and grow larger and less numerous until finally they form two large veins, the *vena cava ascending*, and the *vena cava descending*, which empty into the right auricle. The veins are provided with valves so arranged that the blood can only flow towards the heart. The walls of the veins are not as thick and strong as those of the arteries.

How do veins and arteries differ as to situation?

The veins generally lie near the surface of the body, just under the skin. They may be seen in the backs of the hands and in the temples. The arteries, especially the large ones, are mostly situated far beneath the surface, often running close to the bones and through safe passage-ways. This arrangement possesses the advantage of safety, as there is more danger from injury to an artery than to a vein.

Mention some other differences between veins and arteries.

Arteries carry blood from the heart; veins carry it towards the heart. The walls of the arteries are thicker and firmer than those of the veins. Veins have at intervals valves which prevent the blood from flowing backwards; arteries do not.

What artery carries impure blood? What veins carry pure blood?

It is generally said that arteries carry pure blood, and that veins carry impure blood; but there are two exceptions. The pulmonary artery carries *impure* blood from the right ventricle to the lungs, and the pulmonary veins carry *pure* blood from the lungs to the left auricle.

In case of a wound, how can we tell whether flowing blood is from a vein or an artery?

Blood flowing from a vein is of a dark color, that from an artery is lighter. Blood from a vein flows in a steady stream; from an artery it flows in jets, as the heart beats. The flow of blood is more easily stopped from a vein than from an artery.

Describe the capillaries.

The capillaries are a fine net-work of tubes which connect the arteries and veins. They are so much a part of the arteries and veins that it is impossible to tell where an artery ends, and where a vein begins. It is in the capillaries that the blood gives up the materials for repairing waste and for building new tissues. Here also the blood receives waste matter and particles of worn-out tissue which are to be carried back and cast out of the body through the lungs.

Trace the course of the blood through its complete circulation.

In circulation the blood passes from the right auricle into the right ventricle, from the right ventricle through the pulmonary artery to the lungs, from the lungs through the pulmonary veins to the left auricle, from the left auricle into the left ventricle, from the left ventricle into the arteries, from the arteries into the capillaries, from the

THE ORGANS OF CIRCULATION. 63

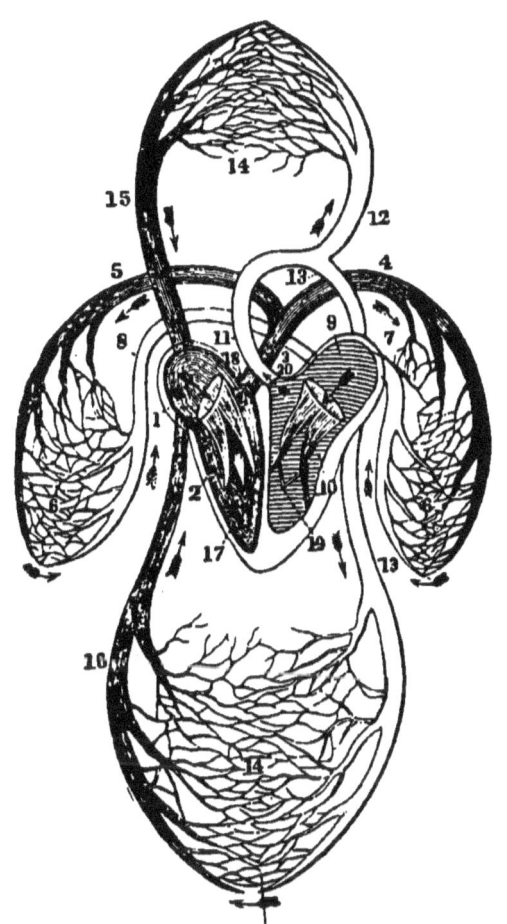

A Diagram of the Circulation.

1, right auricle. 2, right ventricle. 9, left auricle. 10, left ventricle. 4, 5, pulmonary arteries. 7, 8, pulmonary veins. 11, 12, 13, 13, aorta and its branches. 6, 6, pulmonary capillaries. 14, 14, systemic capillaries. 17, tricuspid valves. 19, mitral valves. 18, 20, semi-lunar valves of the pulmonary artery and the aorta.

capillaries into the veins, from the veins back again into the right auricle of the heart, thus completing the circulation.

What is the force which propels the blood in its circulation?

The force which propels the blood through the system is the muscular power exerted by the contraction of the walls of the heart, aided by the elasticity of the arteries and their surrounding tissues.

What is the average rate of the pulsations of the heart?

The average number of heart-beats in an adult man is about seventy-two per minute, but many circumstances will increase or decrease its rate of action. The number varies with age, sex, and individuals. Napoleon's pulse is said to have been only forty. It is not strange to find a natural pulse at one hundred or more. The will, however, has no power either to increase or decrease the number.

In cases of sickness, why does the physician examine the pulse?

Nearly every disease modifies in some respect the condition of the pulse. Consequently its rate, regularity, force, and fulness will, to the trained touch of the physician, be a good index of the physical condition of the patient.

Mention some things which hasten the circulation.

The blood flows more rapidly during digestion, or when the body is under the influence of stimulants, as alcohol, tobacco, or coffee. Mental or physical exercise quickens the action of the heart. Excitement, inflammation, or fever causes the heart to beat more rapidly.

Mention some things which retard the circulation.

The blood flows more slowly during sleep, or when the body is under the influence of narcotics. Tight clothing obstructs the flow of the blood. It is also retarded by fear, or by any cause which produces fainting.

How does the blood keep the body warm?

The natural temperature of the body is about 98° F. summer or winter; this temperature rarely varies more than a degree or two, except in disease. How this heat is maintained is a question not fully understood. We can perhaps explain it partially as follows: The element in coal, or wood, which burns in our stoves and produces heat, is *carbon*, or charcoal. Air enters the stove through the damper. When the fire is kindled, the oxygen of the air combines with the carbon of the coal or wood, and produces heat and light. Whenever carbon and oxygen unite, heat and carbonic acid are produced. If the combination takes place slowly, no light is produced. It is undoubtedly true that our bodies are kept warm by a similar combination of carbon and oxygen. The fuel of the body is the carbon, taken in the form of sugar, starch, and fat. The oxygen is obtained from the air we breathe. The combining, or burning, does not take place in any one part of the body; but is slowly going on in the blood throughout all parts of the body. This makes the blood warm. The circulation helps to equalize the heat throughout the body, as the veins, arteries, etc., are so many pipes containing warm blood which gives up its heat to surrounding tissues.

CHAPTER XIII.

HYGIENE OF ORGANS OF CIRCULATION.

What may be said in regard to the number of diseases which belong to the organs of circulation?

Although constantly in use and easily affected by mental conditions and muscular exercise, the organs of circulation are affected by few diseases. And in many cases where diseases seem to belong to the heart, the real difficulty lies in other organs. The heart is wonderful for its vitality. It is the last organ of the body to part with its vital energy.

Name some things which should be observed in the care of the organs of circulation.

Among the things to be observed in the care of the organs of circulation are the following : Care should be taken to avoid sudden changes in temperature; tight clothing should not be worn ; great excitement, anger, or any violent exercise of mind or body should be avoided ; hemorrhages from wounds, etc., should be guarded against ; there should be no excessive use of tobacco, alcohol, or unnecessary drugs of any kind ; care should be taken to prevent the introduction of poisons, animal or vegetable, through the broken surface of the skin by absorption ; the body should have at all times a proper amount of physical exercise, plenty of sleep, good food, and pure air.

Why should sudden changes of temperature be avoided?

The blood-vessels are contracted by cold, consequently a chill drives the blood to other parts. The chilled parts are weakened, while the other parts are congested, or overburdened with blood. Congestion of the lungs from sudden cold is a frequent and dangerous occurrence. Exposure to intense heat, whether of the sun or of a furnace, causes a sudden prostration known as sun-stroke. The danger lies in the congestion which occurs in the internal organs.

How does tight clothing affect the circulation of the blood?

Compression of any kind hinders the free circulation of the blood. Tight clothing about the waist interferes with the action of the heart and lungs. Tight shoes cause cold feet. The tight dressing of the neck deprives the brain of the necessary amount of blood, and also hinders the free return of impure blood from that organ. This last is an item of great importance to brain-workers, and to persons who are inclined to any brain disease.

Why should excitement of mind, or violent exercise, be guarded against?

Violent physical exercise, like running to catch a train, greatly increases the action of the heart, and sometimes to such an extent that it forces blood into the lungs faster than the lungs can get rid of it. This clogs the lungs, causing difficult breathing, and in some cases, the rupture of a blood-vessel, and dangerous bleeding from the lungs. Excitement, anger, fright, etc., if sudden and intense, frequently check the action of the heart so suddenly as to produce *fainting*, or swooning. The cause is that the brain has been suddenly deprived of its supply of blood. Persons have died, instantly, in a fit of anger.

What are hemorrhages, and how should they be treated?

A hemorrhage is the losing of blood, either from a vein or an artery. If the blood from a wound flows in spurts and is of a bright red color, it is from an artery. If it is dark-colored and flows in a steady stream, it is from a vein. If the bleeding is from an artery, put a tight bandage, or otherwise apply pressure, on the side of the wound towards the heart. If from a vein, apply the pressure on the side of the wound away from the heart.

In what way does tobacco affect the circulation of the blood?

Tobacco is a poison, and its use, especially by the young, is attended with more or less danger. It is probably true that there are few people who use it habitually that do not suffer injury by it. It causes nausea, vomiting, dizziness, and weakness. Its prolonged and excessive use produces an irritability and weakened condition of the heart, known to physicians as the "tobacco heart." When a boy first tries to use tobacco, it makes him very sick. If you should feel of his pulse then, you would find it weak. This means that the poison of the tobacco has partially paralyzed, or weakened the action of the heart. It is said, also, that tobacco injures the blood corpuscles.

How does alcohol affect the circulation of the blood?

Alcohol hastens the circulation of the blood, and causes congestion of the blood-vessels. It increases the work of the heart, and thus exhausts its power. It is said that two ounces of alcohol per day will increase the work of the heart four per cent. Alcohol also softens the muscular fibres of the heart, and weakens it by changing the fibres into fat. It so acts upon the blood corpuscles as to reduce their capacity to absorb oxygen and carry out carbonic acid, etc. The walls of the blood-vessels also are weakened by the use of alcohol, and when from the same cause the blood-vessels of the brain are overcharged, they are liable to break, causing *apoplexy*.

What error is often committed in the use of drugs for "purifying the blood"?

Many people regard the blood as the source of all diseases, and "to purify the blood" is, as they believe, the cure. Consequently *quacks* have been able to make fortunes by selling medicines, recommended for this purpose. The nature of the disease and the proposed remedy should be well understood before the taking of *any* medicine, and especially medicines which profess to cure almost everything by "purifying the blood."

How is there danger of absorbing poisons through the skin?

Poisons are frequently taken into the blood through the skin, especially if there is a break in the cuticle, or outer skin. Persons often poison their hands with common wood-ivy. Painters absorb so much lead through the pores of their hands that they are sometimes attacked with colic. Contagious diseases are sometimes taken in this way. Cosmetics, hair-dyes, etc., are dangerous, because they often contain poison which may be absorbed into the system.

Why does a healthy condition of the blood and organs of circulation require proper exercise?

Exercise of the muscles is not only a means of health to the muscles themselves, but to the entire system. Such exercise produces pressure upon the blood-vessels, and increases the force and rapidity of the circulation, thus promoting the consumption of oxygen by the tissues and the escape of carbonic acid and other waste products.

Why does a healthy circulation require sleep? Good food? Pure air?

All parts of the body share, directly or indirectly, in the benefits of sleep. Repair and waste are going on at all times, whether we are awake or asleep; but during waking hours waste is greater than repair, while during sleep repair is far greater than waste. The heart and organs of circulation are less active during sleep, consequently they share in the benefits of the period of rest and repair.

The materials which compose the blood are all obtained from the food, therefore good blood is dependent upon good food.

We say that the blood is purified by being mingled with the air in the delicate cells of the lungs. The purifying element of the air is its oxygen. If the air lacks oxygen, the blood stagnates, the heart acts slowly, waste matter is not cast off, but remains in the blood and obstructs the whole system.

NOTES AND SUGGESTIONS.

Note 1. Use of Microscope to Examine Corpuscles.—A microscope of moderate power is sufficient to show the corpuscles of the blood. The white corpuscles will not be so readily discovered as the red ones. A drop of blood can be obtained for such an examination by winding a string tightly around the finger and then pricking the finger with a needle. It will be of interest to examine the blood of two or three other animals in this connection. The shape and size of corpuscles will be found to differ in various animals; but in the case of most of the domestic animals, it will be hard for the unskilled observer to distinguish them from those of man.

2. *How to Study Clotting of the Blood.*—This may be done by placing the blood of a sheep, or of an ox, in a deep glass dish. After a little time clots will form and will lie in the centre of a light yellow fluid. The clot contains the globules, fibrine, and most of the coloring matter of the blood. Fibrine may be seen in the fibrous filaments which remain after washing a clot of blood in water. The yellowish fluid contains water, salts, and a little coloring matter.

3. *Circulation in a Frog's Foot.*—The showing of the circulation of blood in the membrane of a frog's foot is an important and interesting object lesson. The frog may be tied up in a wet cloth, leaving

one of his hind-legs outside. By attaching a cotton thread to two of his toes, and by the exercise of a little skill, his foot may be brought and fastened so that the membrane will lie under the object-glass of the microscope.

4. *Dissect a Sheep's Heart.*—A sheep's heart with lungs attached may be had from the market. Its dissection before the class will probably give pupils a better idea of the structure of the heart than can be obtained in any other way. Show pericardium, auricles, ventricles, valves, veins, and arteries. It would be well for the teacher to perform at least one such dissection, before undertaking to do it before the class.

5. *To Find the Pulse.*—Pupils should be taught to find and to count the pulse in their own wrists, also in the wrists of others. Every person should know the rate of his pulse when in a state of health, in order that he may know when an increase or decrease becomes an evidence of disease.

6. *Proofs that the Blood does Circulate.*—People lived for thousands of years before it was discovered that the blood was in motion. When Dr. William Harvey in 1619 announced that it did circulate, he was not believed, and was subjected to ridicule and persecution. But the fact remained, and is now universally known. As proofs, it may be mentioned, that if certain chemical or coloring matter be put into the blood-vessels on one side of an animal, it will in a few seconds appear in the blood-vessels on the other side. In the case of the frog the blood can be *seen* in circulation. The beating of the heart and the motion of the pulse are confirming evidences.

7. *Transfusion of Blood.*—Persons are often prostrated and sometimes brought to the point of death by loss of blood. In such cases blood from the veins of a strong healthy person, or from an animal, is sometimes passed into the veins of the person prostrated. This operation is called *transfusion of blood*. It was first performed about two hundred years ago, and it was then even believed that old people might be restored to the strength of youth by its use. Several deaths occurred from the operation, and it fell into disrepute. During the last few years it has been revived with good results. Better knowledge of the subject, and better surgical appliances, have nearly removed the danger from its use. It will not, however, overcome the effects of old age, but it will in certain cases save life.

8. *Vitality of the Heart.*—The tissues of the heart are remarkable for their vitality. This is especially shown in the case of some animals. The heart of a turtle will pulsate, and its blood will circulate for a week or more after its head is off. The heart may even be taken out of the body, when it will continue to throb regularly for hours. The heart of an alligator, or a frog, or a snake, is likewise remarkable.

9. *Some Animals are Called Cold-Blooded.*—The blood of reptiles is much cooler than that of mammals and birds. The heart of a reptile has only three cavities instead of four, two auricles and one ventricle. Consequently, pure and impure blood are mixed in this ventricle, and the blood is not so perfectly oxidized. The blood of fishes is also cold. Fishes use little oxygen. What they get is obtained from the air in the water. Fishes have hearts with only *two* cavities. The blood is forced by these to the *gills*, which serve as lungs, and from these it is distributed immediately to the different parts of the body.

10. *Ancient Idea of the Heart.*—The ancients thought the heart to be the centre of the affections, the seat of courage, faith, love, and all the other virtues. Modern science has robbed it of its romance, but many words in common use are left to remind us of this fanciful theory. We say in case of grief, "*heart-broken;*" a *heart* is a symbol for the sentiment of love; one person is "*heartless,*" another is "*kind-hearted,*" we learn our lessons "*by heart,*" etc. The word "*courage*" is from *cor*, meaning the heart.

CHAPTER XIV.

ORGANS OF RESPIRATION.

What is meant by respiration?
Respiration is the act of breathing. It consists of two operations—taking in the air, called *inspiration;* and driving out or expelling the air, called *expiration.*
What are the objects of respiration?
The objects of breathing or respiration are, to provide fresh oxygen to be combined with the carbon in the blood in order to keep up the warmth of the body and aid in the assimilation of food; to carry off waste products of the body, as carbonic acid and watery vapor; and, indirectly, to give voice, or the power of speech.
Name the organs of respiration.
The organs of breathing or respiration are the *larynx, trachea,* and the *lungs.* The lungs constitute the chief organ of breathing. The trachea and its branches are simply pipes which make a passage-way for the air. The larynx, situated at the top of the trachea, is really the organ of voice.
Describe the larynx.
The larynx is a small, triangular, cartilaginous box, situated just below the root of the tongue, and at the top of the trachea, or windpipe. The front of the larynx produces the prominence in the neck, sometimes called Adam's apple. The opening from the throat into the larynx is called the *glottis,* and over this opening is a spoon-shaped

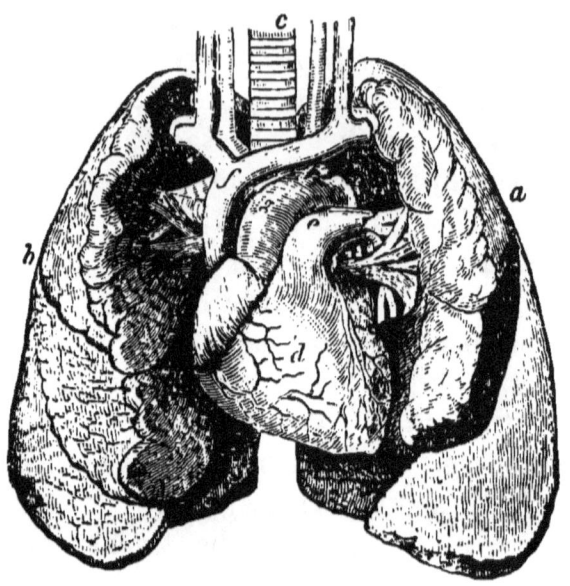

THE LUNGS.

a, the left lung.
b, the right lung.
c, the windpipe.
d, the heart.

e, the great artery carrying blood to the lungs.
f, the great vein.
g, the great artery carrying blood to the body.

lid called the *epiglottis*, which opens when we breathe ; but closes when we try to swallow, so that the food passes over into the œsophagus. In the larynx, attached to its sides, are bands of membrane, called *vocal cords*, by the vibrations of which voice is produced.

Describe the trachea, or windpipe.

The trachea, or windpipe, is a cartilaginous tube extending from the larynx to the lungs. It is four or five inches long and about three-fourths of an inch in diameter. It is composed of rings of cartilage, connected and covered by muscular fibres. These rings serve to keep the tube open for the passage of air. At the lower end the trachea divides into two branches, called *bronchi*, one of which leads to the right lung, and the other to the left lung. As they enter the lungs these branches again divide into smaller tubes, called *bronchial tubes*. These keep on dividing until they are not more than $\frac{1}{100}$ of an inch in diameter. They end in air-cells.

Describe the lungs.

The lungs are two large, pinkish, spongy organs situated in the cavity of the chest, one on each side, with the heart between them. The right lung is larger than the left, and consists of three parts, or *lobes*. The left lung has only two lobes. In the lungs are thousands of very small blood-vessels, so arranged that each one is completely surrounded by air. The oxygen of the air enters the blood by passing through the thin walls of these blood-vessels, and carbonic acid and watery vapor escape from the blood in the same way.

Describe the covering of the lungs.

A double membrane, called the *pleura*, forms a covering for the lungs. One of these membranes is attached to the walls of the chest, and the other to the lungs. These layers of membrane secrete an oily fluid which lubricates them so that they glide upon each other with the utmost ease.

With what are the air-tubes lined?

All the air-tubes, including those of the nose and lungs, are lined with an extremely sensitive and delicate lining, or mucous membrane. This membrane is so sensitive that it will not permit the presence of anything except pure air. A particle of food, a drop of water, or anything offensive in the air, coming in contact with it, irritates it, causing a violent cough, which is the effort of nature trying to expel the trespassing particles.

How does the air reach the larynx?

The air reaches the larynx through the nostrils or through the mouth. The nostrils, however, are the true breathing passages. They are especially fitted to warm, cleanse, and moisten the inhaled air. They open into the upper part of the throat by two openings, similar to those in the nose.

Why is it better in breathing that the air should pass through the nostrils?

The air in passing through the nostrils is warmed and moistened because the passage-way is long and narrow, and the air comes in contact with a large extent of moist, warm surface. Besides this, there are hairs just within the nostrils, and hair-like filaments on other portions of the nasal cavity, which keep out dust and other foreign matter. If we breathe through the mouth, the air passes into the lungs without being warmed, it dries the throat, and is more liable to carry to the lungs dust and other impurities.

CHAPTER XV.

RESPIRATION AND VOICE.

What is the composition and use of the air?

Common air is mainly composed of two gases, oxygen and nitrogen, mixed in about the proportion of one part in volume of oxygen to four parts of nitrogen. It also contains small quantities of carbonic acid and watery vapor. Oxygen is the important element in the air. It enters into the composition of all animal and vegetable matter. The want of a fresh supply of it for five minutes would cause death. Nitrogen is not used either by animals or plants; neither is it injurious to them. Its use is to dilute the oxygen. Carbonic acid is not used by animals, while plants thrive in it. Air containing more than a certain amount of carbonic acid, four or five per cent., acts as a poison to animals. Animals take in oxygen and give off carbonic acid, while plants take in carbonic acid and give off oxygen. Thus each serves the other.

How is the air made to enter the lungs?

The lungs are not muscular, consequently they have no power to act for themselves, but by the action of the muscles of the walls of the chest and the diaphragm (a broad muscular partition between the chest and the abdomen) the cavity of the chest is enlarged. The lungs, consisting largely of little air-cells and air-tubes, are elastic; and when the cavity of the chest expands by the raising of the ribs and by the descent of the diaphragm, the lungs expand also, the air-cells open, and the outside air rushes in to fill them, just as air enters a bellows when its walls are separated.

How is the air expelled from the lungs?

Attached to the ribs is another set of muscles which pull the ribs downward, the diaphragm also is pushed upward against the lungs by the contraction of the muscles of the abdomen. By these movements the cavity of the chest is greatly diminished, and the air is pressed out of the lungs through the bronchi, trachea, and nostrils.

What is the average number of respirations per minute in an adult?

In health, the average number of respirations per minute by an adult is about eighteen. The number is greater in women and children than in men. The number is increased by exercise, food, and stimulants, while it is diminished by inactivity, starvation, and mental depression.

How does breathing purify the blood and nourish the body?

The air taken into the lungs by breathing purifies the blood in two ways, by taking out waste matter and by supplying fresh oxygen. A portion of the oxygen of each breath passes through the walls of the air-cells and of the capillaries in the lungs, and enters the blood. In the blood it joins itself to the red corpuscles and is carried to the heart and out through the arteries to the capillaries in all parts of the body. Here it leaves the corpuscles and becomes a part of the material which helps to nourish and build up surrounding tissues. As the blood discharges its load of oxygen from the capillaries, it takes up at the same place a load of carbonic acid, which it carries back to the lungs.

The blood also brings other waste matter from the capillaries, some of which is discharged through the skin, and some through the kidneys.

What amount of air is required by the lungs?

About twenty cubic inches of air pass in and out of the lungs every time we breathe. If we breathe eighteen times per minute, we use about three hundred and sixty cubic inches of air in that time. This is equivalent to thirteen cubic feet, or seventy-eight gallons, per hour; and three hundred cubic feet, or about sixty barrels, per day.

Describe the vocal cords.

The vocal cords are elastic membranes stretched across the opening to the larynx and attached to the sides. When not in use they spread apart, leaving a narrow opening, through which the air passes into and out of the lungs, without producing voice. The changes in the position of these cords are made by the action of muscles, with which the larynx is provided.

How is sound produced by the vocal cords?

When we wish to use the vocal cords in producing sound, the muscles of the larynx contract and draw the cords tight across the opening. The air passing between the parallel edges of the cords, if the current is sufficiently strong, sets them vibrating, and thus sound is produced in the same way as by the rapid vibrations of the tongues of the accordion, or the strings of a violin.

How are the tones of the voice changed by the vocal cords?

Sounds are varied chiefly by changing the length and tension of the vocal cords. When high tones are to be produced the cords are made short, tight, and are drawn close together; low tones require the opposite conditions. The loudness of a tone depends upon the quantity of air, and the force with which it is expelled from the lungs. The tones of the voice are also modified by the lips, tongue, and teeth.

What is meant by " change of voice "?

In early childhood the character of the voice is nearly the same in both sexes. About the age of fourteen, the vocal organs begin to enlarge rapidly. In a boy, the larynx becomes almost double its former size, and the voice "changes" and takes a masculine tone. In girls the larynx increases about one-third in size. The larynx of a woman is smaller than that of a man. During the period of change, especially with boys, muscular control of the organs is diminished, which produces an uncertainty of tones in singing or declaiming.

Why does a cold sometimes affect the voice?

The voice is often changed and sometimes destroyed by what is known as a *cold*. In such cases the muscles which control the vocal cords become so inflamed, or swollen, that they cannot be used. Hence such a person can only speak in a whisper. A whisper can be produced by the lips, tongue, and teeth, without the aid of the vocal cords.

What actions are modifications of ordinary movements of respiration?

Coughing, sneezing, sighing, yawning, laughing, etc., are caused by a modification of the ordinary movements of respiration. A *cough* is a violent expiration in which the air is driven through the mouth, caused by some irritation of the air-passages. A *sneeze* is a similar act, except that the air is driven through the nose. A *sigh* is a prolonged inspiration, followed by an audible expiration. A *yawn* is very much like a sigh. *Laughing* and *crying* are produced by short, rapid contractions of the diaphragm.

CHAPTER XVI.

HYGIENE OF RESPIRATION.

What can be said of the liability of the organs of respiration to disease?

On account of their constant use, their delicate structure, and their exposure to cold, dust, bad air, disease germs, etc., the lungs and air-passages are especially liable to disease, and great care is necessary to keep them in health.

Mention some things which should be observed in the care of the organs of respiration.

In order to keep the organs of respiration in a healthy condition, we should breathe pure air, which requires well-ventilated rooms, and freedom from stagnant water and decaying animal or vegetable matter; the organs should not be compressed by tight clothing; the pores of the skin should be kept open by proper attention to cleanliness; breathing should be deep and full, so as to bring every part of the lungs into action, and should take place through the nose instead of the mouth; extremes of heat and cold should be avoided; alcohol and the excessive use of tobacco are sometimes a source of disease, and should likewise be avoided.

Why is it necessary to breathe pure air?

As we have already learned, pure air makes pure blood. And the health of all organs of the body is dependent upon pure blood. Air is made impure by carbonic acid and other matter, thrown out through the lungs and pores of the skin; by the decay of animal and vegetable matter; by dust floating in it; by improperly-constructed or uncovered drain-pipes, sewers, etc. Impure air in all these forms should be avoided as far as possible.

How does tight clothing interfere with respiration?

Tight clothing about the neck or chest prevents the free action of the lungs, and interferes with natural breathing. The practice of tight lacing presses upon the ribs so as to change the whole shape of the chest. The stomach and liver, as well as the lungs, are not given room to properly do their work. Tight clothing about the neck interferes with the action of the larynx, and is liable to bring on irritation which may end in bronchitis or consumption.

How does cleanliness of the skin benefit the organs of respiration?

The skin is one of the avenues through which impurities from the body are thrown off. If these impurities are allowed to accumulate upon the skin, they clog the pores, and prevent the escape of waste matter. Extra work then is required of the lungs and other organs, which they may not be able to do properly. Hence we may say that bathing keeps the pores open, promotes excretion, and aids in regulating bodily temperature, and in warding off colds, fevers, and other disorders. It is believed by many medical authorities that proper bathing will to a great extent prevent or cure catarrh and colds in throat or lungs.

Why should we breathe deep and full? Why should we breathe through the nose?

By full deep breathing the blood gets the benefit of more air, and is more thoroughly purified. The lungs, also, are made healthier and stronger by the exercise, and consequently less liable to disease.

The reasons for breathing through the nostrils, instead of the mouth, have been given in answer to the same question on page 76.

Why should extremes of heat and cold, or sudden changes of temperature, be avoided?

By the aid of clothing and artificial heating, man is able to endure a climate subject to wide ranges of temperature. Yet care should be taken to avoid exposure to extremes of temperature. The passing from an overheated room out into extreme cold air is liable to cause congestion of the lungs, or diseases of the throat. The living in overheated rooms during winter will make any person, child or adult, tender and delicate, and especially liable to disease from exposure to cold. Living-rooms, as far as possible, should be kept at a uniform temperature of about 68° or 70° F.

In what way does alcohol injuriously affect respiration?

It is said that so small a part of alcohol as one part to five hundred of the blood will materially check the absorption of oxygen in the lungs. Consequently, unable to take up oxygen, the blood retains its carbonic acid and goes back through the system with the poison which it ought to throw off. Alcohol in excess, also, tends to bring on inflammation of the lung-tissues, and hence lessens the breathing capacity. And there is good medical authority for saying that this action on lung-tissues often leads to a form of consumption known as "alcoholic consumption."

What is the effect of tobacco upon the organs of respiration?

The smoking of tobacco, especially if carried to excess, has often caused diseases of the organs of respiration. The strong, heated smoke irritates the mucous membrane of the mouth and throat, causing what is called "smoker's sore throat." It also prevents the perfect oxidation of the blood and interferes with the assimilation of food. The smoking of cigarettes by boys is extremely injurious. The smoke of the paper is irritating to the lungs, the smoke is more directly inhaled, and the ingredients used in their manufacture are often filthy and harmful.

What is meant by malaria?

The word *malaria* means bad air. Hence malaria is a disease caused by breathing air poisoned by gases, or particles of matter that arise from decaying animal or vegetable matter, from drain-pipes, marshes, etc. It is supposed that little atoms, called *spores*, rise into the air from these sources, and, entering the lungs, poison the blood, producing the disorder from which so many people in some parts of the country suffer.

Name some of the diseases of the organs of respiration.

Among the diseases to which the organs of respiration are subject are: *bronchitis* (brŏn-kī'-tis), an inflammation of the bronchial tubes; *pleurisy*, an inflammation of the pleura, or membranous covering of the lungs; *pneumonia*, an inflammation of the lungs, chiefly affecting the air-cells; *consumption*, a disease which destroys the substance of the lungs; *asphyxia* (as-fix'-ĭ-a), suspended animation from suffocation, drowning, etc.; *diphtheria*, a peculiar sore throat, a contagious, often fatal, and much-dreaded disease; *croup*, an inflammation of the mucous membrane of the larynx and trachea.

NOTES AND SUGGESTIONS.

Note 1. *Structure of Lungs, etc.*—In order to study the structure of the lungs, trachea, larynx, etc., get from the market the lungs of a

HYGIENE OF RESPIRATION.

sheep or pig, with the trachea attached. See that none of its parts are mutilated in cutting. The lungs may be inflated by the use of a bellows. Show structure of trachea, larynx, and vocal cords. The lungs of animals are sometimes called *lights*. Place them on water, they will float.

2. *Experiments with Oxygen and Carbonic Acid.*—If the apparatus can be obtained, some interesting experiments can be made by showing the effects of oxygen and carbonic-acid gas on a burning candle, and also upon some small living animal as a mouse. A text-book on chemistry will give directions for preparing and collecting these gases.

3. *Danger from too Much Oxygen.*—The experiments mentioned above with oxygen would show the effects of an excess of oxygen in the air. It would become a very destructive agent; the tissues of animals would be rapidly consumed, and all substances capable of being set on fire would burn beyond control. In pure oxygen steel would burn, as may be shown by burning a watch-spring in a jar of oxygen.

4. *Show Carbonic Acid and Watery Vapor in the Breath.*—The test of carbonic acid is lime-water. Carbonic acid will form with lime-water a milky precipitate of carbonate of lime. Take a glass tumbler with a little lime-water in it, and breathe, or blow the breath into it, through a glass tube. The lime-water will turn milky, and if allowed to stand the white carbonate will settle to the bottom.

Water vapor, or moisture, in the breath may be shown by breathing against cold glass or steel. The moisture will be condensed on the surface.

5. *How Some Animals Breathe.*—It will be of interest to notice the organs of respiration in some animals; for instance, the frog, fish, whale. Do plants breathe? How?

6. *Black Hole of Calcutta.*—During the English war in India, one hundred and forty-six English prisoners were shut in a room scarcely large enough to hold them. There were two small windows, both on the same side of the room, through which air could enter. At the end of eight hours only twenty-three were alive, and they were in a deplorable condition, from want of air.

CHAPTER XVII.

ORGANS OF MOTION.—MUSCLES.

What are the muscles?

The muscles are the organs by which the movements of the body are performed. They are the deep-red *flesh* of the body, or as it is called in animals, "lean meat." The number of muscles is about five hundred.

What are the uses of the muscles?

The uses of the muscles are, to give the body the power of motion, to help hold the bones in position, and, together with the fat, to fill out the body and give it a symmetrical form. They also shield the blood-vessels; diminish the force of shocks and blows; and in connection with the bones enclose the cavities of the chest, abdomen, and pelvis.

Describe the structure of muscles.

The muscles are made up of fibres, held together by a delicate tissue, called *connective tissue*. These fibres, if examined under the microscope, appear to be composed of still finer fibres, called fibrils. In shape and in length, muscles vary greatly. Some are round; others are flat, square, or triangular. In length, they vary from one-eighth of an inch to nearly three feet. Muscles are large and thick in the middle and small at the ends. The middle part is called the body, or *swell*, of the muscles, and possesses the power of contraction.

How do muscles produce motion?

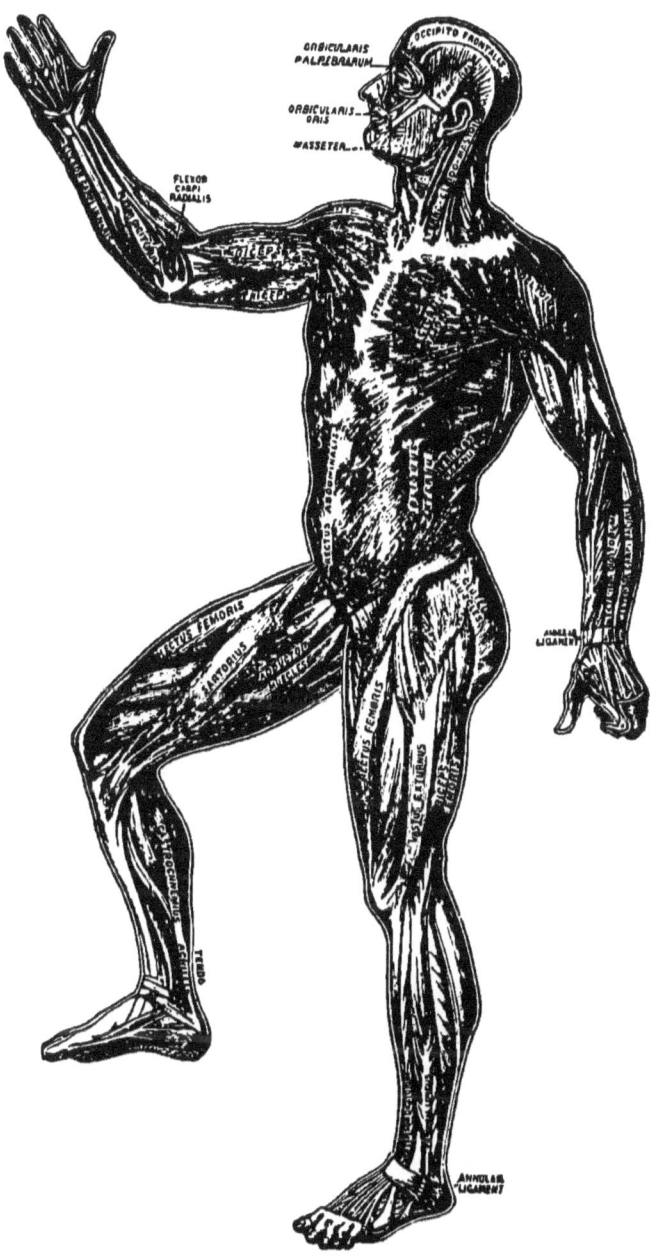

THE MUSCLES.

All the varied and wonderful movements of the body, made in walking, working, talking, breathing, the beating of the heart, etc., are made by the contraction of muscles. That is, the muscles shorten themselves, and bring their two ends nearer together. In this way the bones to which they are attached are moved. Every muscle is provided with nerves which connect it with the brain, and through these nerves it receives the stimulus, or the nerve-force, which gives it the power to act. After a muscle has become shortened, it will remain so for only a short time. It soon becomes tired and begins to relax.

How are muscles mostly arranged?

Muscles are, for the most part, arranged in pairs, or corresponding sets, so that when motion is produced in one direction by one set, another set or group of muscles brings the limb back to its place again. The muscle which bends a joint is called a *flexor* muscle; a muscle which brings a joint back to its place is called an *extensor* muscle.

According to the control which the will has over muscles, how are muscles classified?

According as they are, or are not, under control of the will, muscles are divided into two classes, *voluntary* muscles and *involuntary* muscles. Voluntary muscles are those which are under control of the will; such as the muscles of the hands, arms, legs, trunk, face, etc. Involuntary muscles are those which are not under the control of the will; the muscles of the heart and stomach are involuntary muscles.

What are tendons, or sinews? What is their use?

Tendons, or sinews, are strong, white, inelastic cords, which connect the muscles with the bones. Tendons are easily felt on the inside of the wrist. Children often amuse themselves by taking the leg cut from a fowl, and moving the foot by pulling the tendons. The largest tendon of the body is in the heel, called the tendon of Achilles.

The use of tendons is to give strength and freedom of motion, and at the same time place the moving muscles far enough away to prevent clumsiness. For example, the muscles which open and close the hand are situated in the forearm toward the elbow, and tendons extend from them through the wrist to the fingers. It is easy to see how awkward, clumsy, and inconvenient it would be if, instead, the muscles were placed upon the wrist, or the fingers themselves.

Of what use is "fat" to the body?

Fat usually constitutes about one-twentieth part of the weight of the body. It is situated about the internal organs, between the muscles, under the skin, and about the joints of the body. It acts as cushions to the various structures; it helps to maintain an even temperature in the body; it fills up inequalities, and so improves the symmetry of form; it also serves for nutrition in time of need, especially in emaciating diseases or starvation.

CHAPTER XVIII.

HYGIENE OF MUSCLES.

Name some things to be observed in the care of the muscles.

The health of the muscles requires that the muscles receive a proper amount of exercise. This exercise should be regular, should not be too violent, nor should it be continued too long. The muscles should have rest after exhaustive strain or long-continued exercise. They should be abundantly supplied with pure blood, which requires proper food, pure air, and freedom from compression. They should not be weakened by the use of alcohol.

How is exercise of muscles beneficial?

The beneficial effects of exercise upon the muscles are very marked. By use the muscles become larger, stronger, and more compact. By disuse they become smaller, soft, and weak. Such exercise, however, not only benefits the muscles themselves, but the good effects are shared by all the other organs of the body. The circulation of the blood is more vigorous, the temperature of the body rises, the brain becomes invigorated, the appetite and power of digestion are increased, the skin and lungs do their work more thoroughly, and the whole body thrives under its influence.

How may muscular exercise be injurious to the body?

Exercise is good while it stimulates nature to build up the tissues. If the exercise is too violent, or too long-continued, it is injurious, because it tears down, or wears out, faster than nature can build up. Besides, muscles may be strained or ruptured, and blood-vessels burst by violent exertion. Useless feats of strength and endurance should be avoided.

When is the best time for exercise? Name some varieties of exercise which are beneficial.

The best time to take exercise is, perhaps, about two hours after a meal. It certainly should not be taken just before or immediately after a meal, unless the meal or the exercise be very light. Among the varieties of exercise which are beneficial are : 1st, those that, as nearly as possible, bring into equal action *all* the muscles of the body, as swimming, horseback riding, base and foot ball, lawn tennis, etc.; 2d, those that exert the muscles of the *upper* part of the body, as rowing, bowling, billiards, etc.; 3d, those that serve principally to develop the muscles of the *lower* part of the body, as walking, skating, dancing, bicycle riding, etc.

Why should rest be taken after continued muscular exercise?

A period of repose should follow a period of muscular exertion, in order that new material may be brought to the muscles by the blood to replace that worn out by the exercise. Sore muscles and stiff joints will often be avoided if the exercise is made to cease gradually, not all at once. As repair goes on faster during sleep, we may say that sleep is the best means of restoring vigor to tired muscles.

How do healthy muscles depend upon good blood?

The muscles depend upon the blood, not only for the materials necessary to their growth and nourishment, but for the removal of waste matter which, if allowed to remain, would poison, and so cause disease of the muscles. Both of these things require pure blood, and, as we have already learned, pure blood demands good food and pure air. Moreover, no article of dress should be worn which is tight enough to prevent the free flow of blood to any muscle, as it will interfere with its nourishment.

What is the effect of alcoholic drinks upon the muscles?

Indulgence in beer, wine, or liquors containing alcohol, never does any good and often does harm. The injury may be direct or indirect. Careful experiments have been made which show that when spirits are used the muscles will not act with as much power, nor for as long a time, as when they are not used. Continued use of alcoholic drinks causes a change in the structure of the muscles. The connective tissue and fat in them become too abundant, and take the place of proper muscular substance. The staggering of the drunkard shows that his muscles are not under good control. This probably is due to the fact that the nerves which govern these muscles are partially paralyzed. Mechanical work requiring great skill cannot be well done by a man addicted to the use of alcoholic stimulants. Indirectly the injury to muscles is due to weakened will, impaired digestion, enfeebled heart, or disease of organs whose office it is to carry waste matters away from the body.

Mention some diseases of the muscles.

St. Vitus' Dance is a disease of the voluntary muscles, in which the muscles have an irregular and spasmodic motion beyond the control of the will. *Convulsions* are an involuntary contraction of the muscles; consciousness is wanting, and the limbs are either rigid or in spasmodic action. *Locked-jaw* is a disease characterized by contraction of muscles, accompanied by spasms. It usually begins in the lower jaw. A slight injury or wound is sometimes the cause. *Rheumatism* is a disease which affects mainly the connective, white, and fibrous tissues of the larger joints. There are two forms, called chronic and acute. The former is of long continuance; the latter terminates more speedily.

NOTES AND SUGGESTIONS.

Note 1. *Show the Structure of Muscles.*—The gross structure of muscle may be shown by examining a piece of well-boiled corned-beef. Put the beef on a firm table, and pick it to pieces with two darning-needles. Notice the connective tissue and the larger muscular fibres. When the fibres become too small to manage without, use a magnifying-glass. The small fibres may be examined with a microscope.

The form and structure of muscles can also be readily shown in the hind-leg of a frog. Kill the frog by putting him in a covered jar containing a pint of water to which a little ether has been added. Remove the skin from the hind-leg. Notice the muscles between the knee and ankle joints, also the tendons running to the toes. The nerve of this muscle may be seen, as a slender white thread, entering the muscle just below the knee.

2. *Name and Locate Some Important Muscles.*—Pupils should be taught to name and locate some of the most important muscles of the body. This can be done by reference to the cuts in some text-book, or on some chart. They should also be able as far as possible to point them out on their own persons, and to tell the office of each.

3. *Show Tendons.*—The leg of a chicken or turkey affords an excellent opportunity for examining tendons. The muscles which move the toes are in that part of the leg known as the "drumstick." After the lower-leg has been cut off, by pulling some of the tendons the toes are closed, by pulling others they are opened.

4. *Tendon of Achilles.*—This tendon was so named because, as the story goes in fable, when Achilles was an infant his mother held him by the heel while she dipped him in the River Styx, the waters of which had the power of rendering any one invulnerable to weapons. His heel did not get wet, and was, therefore, his weak point, to which Paris aimed the fatal arrow.

5. *Animals that Hibernate.*—Some animals, the common tortoise for instance, burrow themselves in the earth in the latter part of autumn, and do not reappear until the opening of spring. Some species of bears become very fat through the summer and fall, but pass the winter in a cave or hollow tree in a state of inactivity, eating no food, so far as is known. The fat, stored up during the summer and autumn, keeps the body warm, and supplies the little nourishment which such a state of inactivity requires. In the spring they come out of their hiding-places lank and hungry.

6. *Strength of Insects.*—Insects, when their size is taken into consideration, present some remarkable examples of muscular strength. A flea harnessed will draw seventy or eighty times its own weight. A horse can draw about six times his weight. A flea can jump several feet at a leap. A common beetle weighing fifteen grains has been known to carry on his back 4,759 grains, or nearly 320 times his own weight.

CHAPTER XIX.

THE NERVOUS SYSTEM.

What is the office of the nervous system?
The office of the nervous system is to give the body intelligence, sensation (five senses), and the power of motion. By intelligence we mean all the higher powers of mind by which man is distinguished from the lower animals ; sensation includes feeling, seeing, hearing, tasting, and smelling ; power of motion may include all the involuntary motions of the vital organs, heart, lungs, stomach, etc., as well as those made by authority of the will, for all muscles, voluntary and involuntary, depend upon the nervous system for their stimulus to action.

Name the organs of the nervous system.
The organs of the nervous system are the brain, spinal cord, and nerves.

What is meant by nerve-tissue?
Nerve-tissue is the soft, marrow-like substance of which the principal bulk of the brain, spinal cord, and nerves is formed.

What are nerves?
Nerves are simply portions of the brain and spinal cord, which extend to every part of the body. When seen by the naked eye, a nerve looks like a silvery-white cord. With a microscope it is seen to consist of a bundle of delicate little fibres.

Describe the location, size, and shape of the brain.

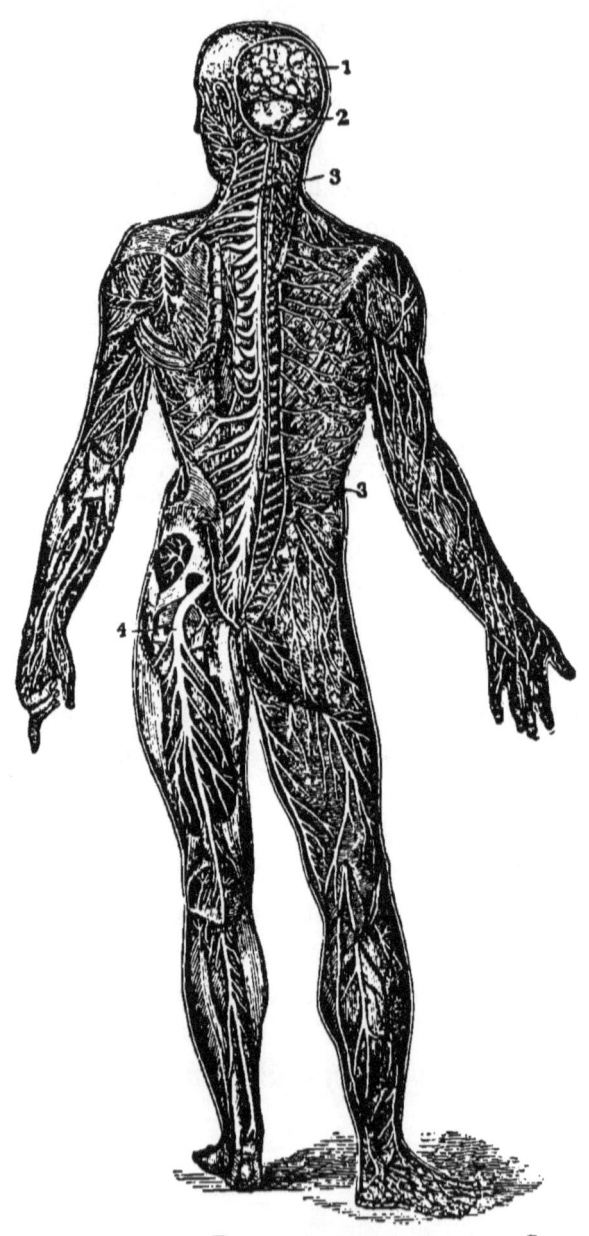

A Representation of the Brain, Spinal Cord, and Spinal Nerves.

1, the cerebrum.
2, the cerebellum.

3, 3, spinal cord.
4, the sciatic nerve.

The brain is the principal organ of the nervous system. It occupies the entire cavity of the skull, and consists of several separate masses of nerve-tissue abundantly supplied with blood-vessels. In shape it is oval, with one extremity larger than the other. The average weight of the brain is about fifty ounces, or a little more than three pounds.

Name the three principal masses which compose the brain.

The three principal masses which compose the brain are the *cerebrum*, or brain proper; the *cerebellum*, or lesser brain; and the *medulla oblongata*.

Of what two kinds of matter is the substance of the brain composed?

The substance of the brain is composed of two kinds of matter,—gray matter and white matter. The outside of the brain is formed of the gray matter, and the inside of the white matter. It is supposed that the gray matter is the generator of nerve-force, and that the white matter is the conductor of nerve-force. What nerve-force is we do not know. It is like electricity in some respects, in others quite unlike it.

Name the membranes which envelop the brain.

Within the skull the brain is enveloped in three membranes, or coverings. They are called *dura mater* (tough covering), the *arachnoid* (spider's web), and *pia mater* (soft covering). The office of the dura mater is to protect the brain and to assist in keeping it together. The arachnoid secretes a lubricating fluid which moistens the surface of the brain. The pia mater contains blood-vessels which help to supply the brain with blood.

Describe the cerebrum, or brain proper.

The cerebrum is that part of the brain which fills the entire upper part of the skull. Its surface has a peculiar wrinkled, or folded-up, appearance; and it is divided into two parts, or halves, by a deep cleft which extends from front to back. Looking at the cerebrum from the top its appearance is not unlike that of the meat of an English walnut. The cerebrum is believed to be the chief organ of the mind, the seat of intelligence and of intellectual processes.

Describe the cerebellum.

The cerebellum, or little brain, is less than one-eighth the size of the cerebrum. It is situated directly under the back part of the cerebrum, and is also divided into hemispheres, and is composed of gray and white matter. It is believed that this part of the brain is the regulator of muscular action.

Describe the medulla oblongata.

The name, medulla oblongata, means "oblong pith, or marrow." This part of the brain is the enlarged upper part of the spinal cord, and forms the connection between the cerebellum and the spinal cord. It is only about an inch and a half long, but it is an important part of the brain. It is believed that the medulla oblongata is the seat of sensation, and that it has control of the action of the heart and lungs. It is probably the most sensitive portion of the body, a slight injury causing death.

Describe the spinal cord.

The spinal cord is an extension of the substance of the brain, and extends downward through the cavity of the backbone. Like the brain, it consists of two kinds of matter, gray and white, but the gray matter is on the inside and the white outside. Between the bones of the spine, the spinal cord sends out to various parts of the body branches called *spinal nerves*. From one set of these branches, the spinal cord receives impressions from various parts of the body and carries them to the brain; through another set, it sends out the commands of the brain to the voluntary muscles.

From their origin, how are nerves classified?

From their origin, or place from whence they start, nerves are classified as *cranial* nerves and *spinal* nerves. The cranial nerves start from the base of the brain, within the skull, or cranium. They consist of twelve pairs, and extend to the eye, ear, tongue, nose, throat, stomach, heart, etc. The spinal nerves, as has already been said, are branches of the spinal cord. They consist of thirty-one pairs, and extend to the arms, chest, abdomen, legs, etc.

According to office, how are nerves classified?

From their office, nerves are classified as *sensory* nerves and *motor* nerves. The sensory nerves are those which carry impressions received through the organs of sense, hearing, sight, touch, taste, etc., from the outside of the body to the brain. The motor nerves are nerves connected with the muscles. They carry the commands of the brain to the muscles, so that the muscles move whenever the brain wills them to.

To what may the nervous system be compared?

The nervous system may be compared to a telegraphic system. The brain is the central office, from which messages are sent back and forth over the nerves, which correspond to the wires. If a nerve, or a wire, leading to any part be severed, communication with that part instantly ceases.

CHAPTER XX.

HYGIENE OF THE NERVOUS SYSTEM.

What may be said regarding the liability to disease of the nervous system ?

Although the nervous system is so delicate in its structure and so mysterious in its work, it is not in itself especially liable to disease. But it is so closely connected with other parts of the body that anything which injures them is liable to injure it. The man whose muscles and digestive, circulatory and respiratory organs are in a healthy condition, usually possesses a healthy nervous system and a vigorous mind. The health of these organs has been quite fully discussed in other chapters, so that little need be repeated here. Something more might be well said, perhaps, regarding the need of sleep, exercise, rest, recreation, and the effects of alcohol and tobacco.

What may be said regarding the effects of sleep on the nervous system ?

The tissues of the nervous system, like those of other organs, depend upon the blood for the materials of growth and repair. Action on the part of the nervous system destroys, or wears out, nerve-tissue. Every thought, sensation, or effort of the will which goes out from the brain consumes a part of its substance. Consequently there must be periods when the brain must stop work for repairs. Whenever we are awake the brain is more or less active.

Sleep is the only period of complete rest for growth and repair. The amount of sleep necessary will depend upon age, health, occupation, and natural temperament of the individual; but sound sleep, and plenty of it, is one great condition necessary to a healthy nervous system.

How does exercise benefit the nervous system?

Physical exercise has already been shown to be beneficial to all organs, because it promotes healthy action and a more vigorous circulation of the blood; but regular and systematic mental exercise is essential to the health of nerve-tissue. Idle minds, like idle muscles, become weak. Cultivation and systematic exercise quicken the action of the brain, and also give it the power of sustaining action for a longer period of time. Mental exercise should also be adapted to the age and health of the individual. Parents and ambitious teachers should see to it that young minds are not overworked.

Why does the brain need rest and recreation?

The confining of the mind for a long time to one subject of thought is exhausting. In order to keep the powers of the brain vigorous and lasting, the lines of mental work should be occasionally changed, and in this way the brain obtains rest. Every one is benefited by a vacation, or a change of employment, in which the ordinary routine of every-day life is broken up. This applies to business men, students, farmers, and mechanics alike.

What are some of the abuses of the nervous system?

The nervous system is sometimes abused by overwork, but fretting and worry are more fruitful causes of mental disorder. Pupils fret and worry about examinations and promotions, and become peevish and nervous, and lay the foundations of future ill-health. Older persons worry themselves into nervous prostration over business cares and the struggle for wealth and position. The nervous system is frequently abused by the use of drugs and medicines for producing sleep, or deadening pain. The disturbance to the nervous system is more than enough to counterbalance all good results. The use of laudanum, morphine, chloroform, for such purposes, except by the advice of a physician, is dangerous. The dose must be constantly increased in order to produce the same effects, and frequently the user is unconsciously led into a terrible and fatal bondage. The injury done to helpless infants by the careless use of "soothing-syrups" and the like is pitiful. An overdose is liable to be fatal, and at best they paralyze the nerves, interfere with digestion, and poison the blood. The use of alcohol and tobacco has, in some respects, a similar effect.

How does alcohol affect the nervous system?

The injurious effects of alcohol upon the nervous system are more marked than upon any other organs of the body. It injures the substance of the brain and impairs the thought-producing power. It sometimes collects in the brain, causing paralysis and death. It destroys a man's judgment and his will-power, leading him to do deeds of crime and violence, and rendering him an utter slave to his animal nature. Its excessive use leads to insanity and to a terrible form of delirium, known as *delirium tremens*.

This question will be more fully answered in a special chapter on Stimulants and Narcotics.

What is the effect of tobacco on the nervous system?

The use of tobacco has a peculiarly injurious effect upon young and growing persons. And adults are by no means exempt from harm, although, if used by them in moderation, it may apparently do no harm except as small doses of any narcotic may do harm. In excessive quantities, its harmful action to old or young is marked. It deranges the working of the brain and nerves so that ideas lack clearness, will-power becomes weakened, memory impaired, and the nerves deprived of the power to steadily control the muscles. Its habitual use is foolish, costly, and ill-becoming, and is a thing to be avoided by a young man who wishes to succeed in life and to keep a constitution strong and vigorous.

NOTES AND SUGGESTIONS.

Note 1. *Difficult to Dissect the Brain of an Animal.*—It is a difficult matter to dissect the brain and spinal cord of an animal, and, unless skilfully done, the attempt to do it before a class will be of little value. If skilfully performed the operation can be made a valuable object-lesson. To show the brain, a sheep's or calf's head may be obtained from the butcher. The spinal cord can be better shown by dissecting the body of a cat or rabbit. The cutting of bones can be mostly done with strong short-bladed shears. Great patience is needed to make such dissections successful.

2. *Nervous System Belongs to Animals Only.*—Organs of circulation, digestion, and respiration belong in modified forms to the vegetable kingdom, as well as to the animal; while a nervous system, so far as is known, belongs to animals only, and the extent of its development determines the position in the scale of intelligence to which the various animals belong. The average weight of the human brain in proportion to that of the entire body is about 1 to 36. The average in mammals is 1 to 186.

3. *Large Brains.*—There have been few men of distinguished ability whose brains have been small. As a rule, the size of the brain is an index of mental development. The following are the brain weights of some distinguished men: Cuvier, 64.5 oz.; Daniel Webster, 53.5 oz.; Abercrombie, 63 oz.; Agassiz, 53.4 oz.

4. *Speed of Thought.*—The rate of motion with which a message travels along a nerve to or from the brain, is thought to be about two hundred feet per second. The estimate has been reached by experi-

ments upon lower animals. It is slower than electricity, but considering the short distance which such messages travel it is practically instantaneous.

5. *Pain Felt in a Limb after Amputation.*—For a long time after a limb has been amputated, pain will be felt in it as if it still formed a part of the body. The explanation is this: Pain, strictly speaking, is in the mind, not in the various parts of the body. The mind only can feel; hence when any nerve brings into the brain the news of an injury, the brain at once refers the pain to that part of the body where the nerve terminated. So in the case of injury to the stump of an amputated limb, the mind refers the pain to the point to which the nerve formerly led.

6. *Peculiarities of the Brain.*—In appearance the brain is the least solid and the most unsubstantial looking organ in the body. Eighty per cent. of its substance is water, nine per cent. is albumen, and the rest fat and small quantities of a few other substances. While the brain is the seat of all sensation, it has no sensation. The spinal cord and every nerve is sensitive to the slightest touch, but the brain may be injured, even cut or burned, without pain.

CHAPTER XXI.

THE SKIN, AND ORGANS OF SPECIAL SENSE.

What are the offices of the skin?

The offices of the skin are quite varied. It makes a covering and a protection for the outer surface of the body; it is an organ of feeling; it helps to provide the means of escape for waste matter; it is a regulator of temperature; and it is an organ of absorption. The skin is an organ of more importance than most people believe it to be.

Describe the structure of the skin.

The skin consists of two distinct layers. The inner layer is called the *dermis*, or true skin; the outer one, the *cuticle*, or scarf skin. The inner layer is also sometimes called the *cutis*, and the outer one the *epidermis*. These two layers are closely united, but they may be separated. This separation takes place whenever a blister is formed; the epidermis, or cuticle, is lifted up from the dermis, or true skin, and a watery fluid collects between.

Describe the cuticle, or scarf skin.

The cuticle in most parts of the body is thinner than the true skin; it is tough, elastic, without feeling, and does not bleed when cut; it is composed of minute flat cells arranged one upon another in layers. The outer layers of these cells, on account of exposure to atmosphere, friction, etc., become hard, and being constantly worn out fall from the body in the form of very fine scales. In the scalp these scales are called "dandruff." As fast as they wear out and fall off, they are renewed from beneath. The material out of which the new cells are formed is supplied by the blood-vessels which lie in the true skin.

Describe the dermis, or true skin.

The dermis, or true skin, lies beneath the cuticle. It is firm, elastic, sensitive, and well supplied with blood-vessels. The outer surface of the true skin rises into little ridges called "papillæ," into which capillaries and nerves are distributed. These papillæ are most numerous where the sense of touch is most acute, as on the tips of the fingers. The true skin contains the sweat and oil-glands, nerves, and hair follicles (little sacs, or pouches).

Of what use are the sweat-glands?

The sweat-glands consist of very fine tubes, coiled up into knots, from each of which a little duct, or pore, extends to the surface of the skin. These glands gather up from the blood in the capillaries perspiration and waste matter, which passes out through the pores to the surface of the body. This work is going on at all times, but the amount of perspiration varies greatly at different times. Anything which tends to heat the body causes it to flow more freely. If the moisture dries up so as not to show itself upon the skin, it is called *insensible perspiration*. When it is poured out faster than it evaporates, and collects in little drops on the surface of the skin, it is called *sensible perspiration*. It is said that there are about two and a half millions of sweat-glands in the surface of the body. The average amount of perspiration per day is about two pints, about two per cent. of which is solid matter. The use of perspiration, besides taking water and worn-out matter from the blood, is to regulate the temperature of the body.

That is to say, evaporation diminishes temperature. When this moisture comes to the surface it evaporates ; the heat which changes it into vapor is taken from the body, which makes the body cooler.

Describe the oil-glands. Of what use are they?

The oil-glands are little glands which pour an oily fluid into the sides of the pits in which the roots of hairs are contained. This oily fluid keeps the hair glossy and the skin soft. It also keeps the skin from absorbing water or other substances too rapidly, and from drying up too fast on a hot dry day.

To what is the complexion, or color of the skin, due?

In the cells on the under side of the cuticle little grains of pigment, or coloring matter, are found. The varying tint of this coloring matter in different persons makes the difference in complexion between the blonde and brunette, the European and the African. The action of the sun on this coloring causes " tan " and freckles.

What are hair and nails, and how do they grow? What are their uses?

The hair and nails are modified forms of the cuticle. They both grow by the addition of new cells at their roots, or lower ends. Hair is found on nearly all parts of the body except the palms of the hands and soles of the feet. Its chief use is to protect the body from excessive heat and cold. The nails provide a kind of armor to protect the ends of the fingers and toes, which are liable to injury.

CHAPTER XXII.

THE SKIN, AND ORGANS OF SPECIAL SENSE.—CONTINUED.

How many special senses have we, and what are they?

It is said that there are five ways by which the brain may receive impressions from objects outside of the body; that is, five gateways of knowledge, called the five senses,—*touch, taste, smell, hearing,* and *sight*.

Describe the sense of touch and feeling.

The skin is the chief organ of the sense of feeling, in which many sensory nerves have their extremities. The impressions received from touch, heat and cold, and pain, are transmitted from the skin through these nerves to the brain; and the brain locates, with more or less accuracy, the part of the skin from which the impressions come. The sense of touch is most delicate on the tip of the tongue, the edges of the lips, and the ends of the fingers. It is least delicate in the middle of the back.

Describe the sense of taste.

The sense of taste is located in the mucous membrane on the upper side of the tongue and the under side of the soft palate. On these membranes there are innumerable elevations, or papillæ, containing the endings of nerve-fibres. Dissolved particles of substances tasted come in contact with these nerve-fibres, and the impression is carried to the brain. The use of taste was originally, perhaps, to guide in the selection of food, but force of habit and highly seasoned foods have greatly impaired its use for this purpose.

Describe the sense of smell.

The sense of smell is very closely connected with that of taste, and we often fail to distinguish between them. The sense of smell has its location in the mucous membrane which lines the upper portion of the cavities of the nostrils. The olfactory nerves are distributed over the surface of this

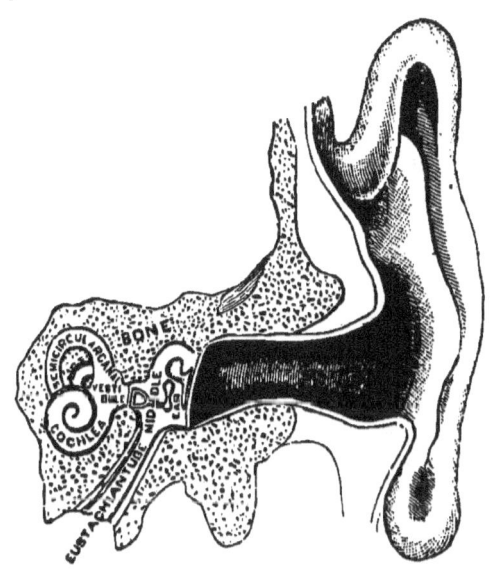

THE EAR.

membrane, where particles of matter which produce odors come in contact with them, and they transmit the impression to the brain. The uses of the sense of smell are to guide us in the selection of food, and to give us warning against bad air.

Describe the ear, or organ of hearing.

The ear, or organ of hearing, consists of three parts—the external ear, the middle ear or *tympanum* (drum), and the internal ear or *labyrinth*. The external ear consists of cartilage, curiously folded for catching sound. Leading inward is the auditory canal about an inch long, across the lower end of which is stretched the membrane of the tympanum, which is kept moist and soft by a fluid wax called ear-wax. The tympanum, or middle ear, consists of a cavity filled with air, and communicating with the throat by a small tube called the *Eustachian tube*. Across this cavity, from the membrane of the tympanum to the inner ear, hangs a chain of three curious little bones, called from their shape the *hammer*, the *anvil*, and the *stirrup*. The internal ear, or labyrinth, consists of winding passages in solid bone. Spread over these passages, like a lining, is the *auditory nerve*, and a watery liquid fills the remaining space. The auditory nerve extends from this part of the ear to the brain.

Explain how we hear.

All things which produce sound vibrate in so doing, and these vibrations are communicated to the air around them. The vibrating air strikes against the membrane of the tympanum, and the vibrations are forwarded to the inner ear, partly by the air in the middle ear and partly by the chain of little bones. In this way the motion is communicated to the liquids in the labyrinth, which in turn excite the ends of the auditory nerve. The auditory nerve conveys the impression to the brain, and the sensation of sound is the result.

Describe the position, shape, and size of the eye, or organ of sight.

The eye is situated in the upper part of the front of the skull, protected by the surrounding bones. The eyeball, which is about an inch in diameter and nearly spherical, rests in a soft elastic bed of fat, which supports and protects it, and at the same time allows it to move freely in all directions.

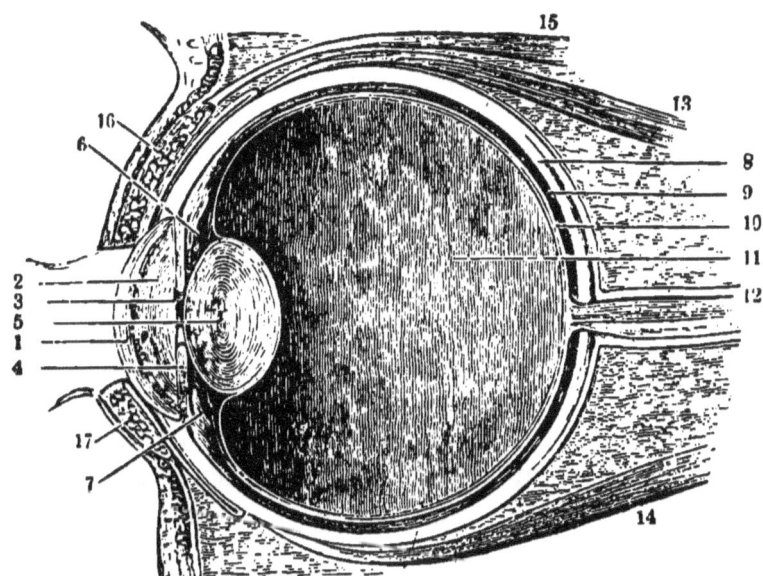

A VERTICAL SECTION THROUGH THE MIDDLE OF THE EYEBALL.

1, cornea.
2, aqueous humor.
3, pupil.
4, iris
5, lens.
6, ciliary processes.
7, canal around the lens.
8, sclerotic coat.
9, choroid.
10, retina.
11, vitreous humor.
12, optic nerve.
13, 14, 15, muscles of the eyeball and of the eyelid.
16, 17, eyelids.

Describe the coatings of the eyeball.

The walls of the eyeball consist of three distinct coats or coverings,—the *sclerotic*, the *choroid*, and the *retina*. The *sclerotic* is the outer covering, and is generally called the "white of the eye." It is tough and strong so as to protect and to keep in shape the delicate structures within, and to afford an attachment for the muscles which move the eye. It has an opening in front in which a round transparent plate, or window, called the cornea, is placed. The *choroid coat* is immediately under the sclerotic. It is much more delicate in structure, and contains nerves and blood-vessels. It is lined with a coating of black pigment which serves to keep out the unnecessary rays of light. The front part of this coat is a movable curtain, called the *iris*, in the centre of which is the opening called the *pupil*. The color of the choroid as seen in the iris gives the color to the eye—as black, blue, gray, etc. The iris is provided with muscular fibres which enable it to move so as to enlarge or contract the pupil for the purpose of admitting more or less light. The *retina* is the innermost coat of the eyeball. It is a delicate transparent film, and lines only the rear half of the eyeball. It is the immediate seat of sight, and is really an extension and an expansion of the optic nerve.

Of what is the interior of the eyeball composed?

The interior of the eyeball is filled with transparent liquid or jelly-like matters, called *humors*. These liquids, or humors, are three in number,—the *crystalline lens*, just behind the iris; the *aqueous* (watery) *humor*, between the crystalline lens and the cornea; and the *vitreous* (glassy) *humor*, behind the crystalline lens and filling the entire back part of the cavity of the eyeball.

What are the uses of the humors of the eye?

The humors of the eye fill up the cavities of the eyeball, and keep the different parts in position and at proper distances from each other. They also receive the rays of light which enter the eye, and so bend and direct them that all rays which start from one point outside of the eye are brought to a focus, or to one point again, on the retina. This, especially, is the work of the crystalline lens. An exact picture of the thing looked at is thus formed on the delicate retina in the back part of the eye. The impression received is carried to the brain through the optic nerve, and the sensation of sight is produced.

Of what use are tears ?

Tears are a special fluid provided by nature for the protection of the eyes. They keep the surface of the eye clean and moist. The lachrymal glands, or tear-glands, are situated in the upper and outer sockets of the eyes, just above the balls. The tears after passing over the eyeball finally find their way to the inner corner of the eye, and there enter little openings which lead to the nostrils, whose inner surfaces they moisten.

CHAPTER XXIII.

HYGIENE OF SKIN AND ORGANS OF SPECIAL SENSE.

Name some things to be observed in the care of the skin.

The skin should be kept at a proper temperature by suitable clothing; it should be kept clean by frequent bathing; its healthy action should not be interfered with by the use of cosmetics and hair-dyes.

How does proper clothing serve to keep the skin in a healthy condition?

Proper clothing serves to keep the skin in a healthy condition by keeping it at an even temperature. It is our defence against the frequent changes of weather in our fitful climate. If clothing is insufficient, the skin is liable to become chilled, perspiration to be checked, and a "cold" is the result. Clothing should be adapted to the climate and the season of the year. There is more danger from too little than from too much clothing. Many children die every year from lung-diseases because of insufficient clothing.

Why should the skin be bathed frequently?

Frequent bathing is necessary in order to keep the pores of the skin open, and also to remove impurities which, if left on the skin, might be absorbed and carried back to poison the blood. Bathing not only cleanses the skin, but if it is done at the proper time, with water at the right temperature, it strengthens and invigorates the whole body. The kind of bathing should be determined somewhat by the state of the weather, the age, health, and occupation of the individual.

Why are cosmetics and hair-dyes objectionable?

The use of cosmetics for the complexion is a frequent source of disease, for most of them contain materials highly injurious to the skin. Lead is very often used as a coloring matter, and is absorbed by the skin, producing ill health by lead-poisoning. Most hair-dyes contain lead, and are objectionable for that reason. They also injure the vitality of the hair, and seldom improve its appearance.

What action has the use of alcohol upon the skin?

Alcohol, taken in beverages of any kind, causes more blood to flow to the skin. This is shown by the flushed face of the man who has been "drinking." If the drinking is excessive and long continued, the redness becomes permanent. This congestion interferes with the proper nourishment of the skin, and the epidermis collects in spots, giving the face the blotchy and degraded look peculiar to the chronic hard-drinker.

What are corns?

Corns are thickened portions of the cuticle or epidermis, caused by pressure or friction. They are usually found on the feet, the result of tight or ill-fitting shoes. They are painful because the hardened cuticle irritates the sensitive cutis below. The extent of the suffering which they inflict upon humanity is a strong warning against tight shoes.

Mention some ways in which deafness is caused.

Deafness may be caused in several ways: by disease of the auditory nerve, the labyrinth, or the tympanum; by the stoppage of the auditory canal by wax or some foreign substance; by a blow upon the ear, or by the concussion produced by an explosion, or by the firing of a cannon near by;

by a cold which settles in the throat and causes inflammation and swelling of the membrane which lines the Eustachian tube, or of other parts connected with the middle or internal ear. If the cause is disease of the auditory nerve or of the internal ear, the deafness may be incurable ; otherwise medical aid, patience, and care will usually effect a cure.

Mention some things to be observed in the care of the eyes.

The eye is a very delicate organ, and great care should be taken to keep it in as perfect a condition as possible. The eyes should never be strained by trying to read or sew with an imperfect light. They should not be fatigued by reading newspapers or other fine print, while riding in a carriage, street-car, or railway-train. It is injurious to look or gaze at objects which are intensely bright,—at the sun, an electric light, a full gas-jet, etc. Books printed in too small type should be avoided. It is not well to read while lying in bed or on a sofa, as some of the muscles of the eye are strained by so doing.

What should be done if the eyes are near-sighted, or far-sighted ?

In case of near-sightedness or far-sightedness, an oculist should be consulted, and the defect, as far as possible, should be remedied by using proper glasses. In a case of near-sightedness the rays of light come to a focus in front of the retina. In a case of far-sightedness they come to a focus behind the retina. A near-sighted person should wear concave glasses, and a far-sighted person should wear convex glasses.

NOTES AND SUGGESTIONS.

Note 1. *To Examine the Structure of the Eye.*—The structure of the eye can be readily examined by dissecting the frozen eye of an ox. It can be sawed from the skull together with a portion of the bone around it, and then frozen by placing it in salt and ice in summer, and in the snow over night in winter.

2. *The Muscular Sense.*—We are said to possess five senses, but there is another, commonly called the *muscular sense*. By this sense we are able to judge of the weight of different bodies by the muscular effort required to lift them.

3. *Senses of Animals.*—The senses of many animals are more acute than those of man. Some birds surpass in sight, the dog has a keener sense of smell, the deer or the rabbit a quicker ear, and the bat is more sensitive to touch.

4. *Color-Blindness.*—Some persons are unable to distinguish between certain colors. This defect is sometimes caused by sickness, but usually exists at birth. It is a matter of considerable practical importance to persons who work on railroads and boats where colored signals are used. Women are seldom color-blind.

CHAPTER XXIV.

STIMULANTS AND NARCOTICS.

What is a poison?

A poison is any substance whose natural effect, when applied externally, or when taken into the stomach or blood, is to produce disease or death.

Into what two classes are poisons generally divided? Define each.

Poisons are generally divided into two classes, *irritant* poisons and *narcotic* poisons. An irritant poison is a poison which irritates, or inflames, the organs of the body with which it comes in contact. A narcotic poison is one which produces sleep, stupor, or paralysis of the nervous system. Some poisons are both irritant and narcotic in their effects.

Name some irritant poisons.

Arsenic, corrosive sublimate, strong acids, potash, lunar caustic, phosphorus, poison ivy, sugar of lead, etc., are irritant poisons.

Name some narcotic poisons.

Laudanum, morphine, chloroform, belladonna, aconite, alcohol, and tobacco are narcotic poisons.

Among those substances which may be said to be both irritant and narcotic in their effects are pink-root, ergot, lobelia, etc.

What is meant by a stimulant?

The word *stimulant* means that which excites, or goads

on. Hence a stimulant is any substance which excites some organ of the body to greater activity without affording it proportionate nourishment. Some poisons are both stimulants and narcotics in their action; but just where the stimulant action ends, and the narcotic action begins, is not easy to determine. Opium, alcohol, hasheesh, and tobacco are substances of this kind.

What is opium?

Opium is the thickened juice of the poppy-plant, which is largely grown in India. The active principle in opium is morphine, which gives the sleep-producing property to laudanum, "soothing syrups," paregoric, Dover's powders, etc. Taken in too large doses it produces a sleep from which there is no waking. Laudanum is made by dissolving opium in alcohol.

What are the effects of opium?

Properly used, opium is one of the most valuable drugs at the disposal of the physician; improperly used, it becomes one of the most harmful. The first effect of opium is stimulating, but the external effects are not as noticeable as are those from the use of alcohol. It is generally first used for the purpose of obtaining relief from pain, but the sense of relief is often so enchanting that the sufferer is unable to resist the temptation to repeat its use. In this way the opium habit, so called, is formed. This habit, if once established, is almost irresistible, and, if continued, is sure to prove fatal. Most physicians now realize the danger, and prescribe opium with great caution. It should never be taken, or given to others, except on the order of a physician, and then exactly as ordered. Many a child has been killed by the use of paregoric, "soothing syrups," and the like, given by ignorant parents or careless nurses.

What diseased conditions result from the continued use of opium?

The continued frequent use of opium causes a diseased condition of the stomach and other digestive organs. The opium-eater is therefore a lean, yellow, sallow person. It also impairs the muscular and mental powers, especially weakening the will. While the intoxication produced by opium is of the most exquisite kind, the after effects are proportionately horrible. The habit can be broken only with intense suffering on the part of its victim.

What other drugs are similar to opium in their usefulness and in their liability to do harm?

Chloral, chloroform, ether, and cocaine are similar to opium in their action upon the nerves and brain in diminishing sensibility to pain and in producing sleep. In the hands of careful physicians and skilful surgeons they have proved to be great blessings to suffering humanity; but used without proper medical advice, they are dangerous agents,—no longer remedies, but poisons. Like opium, they are seductive in their nature, and their use frequently leads to a habit so strong that few possess the will-power sufficient to throw it off.

What is alcohol?

Alcohol is a clear, water-like liquid, possessing a hot, biting taste, obtained by distilling fermented vegetable juices. It exists in all fermented liquors to a greater or less extent. Distillation does not make the alcohol, it is only a process by which the alcohol is separated from the rest of the liquid. It is alcohol which gives the intoxicating property to brandy, whiskey, gin, rum, etc. Alcohol, it is said, was first distilled from wine by an Arabian chemist about the year 1000 A. D. The word "alcohol" is of Arabic derivation.

What is fermentation?

If the juices of ripe fruits of any kind, as of apples or grapes, are allowed to stand in a warm place for a time, they begin to "work," or ferment. New cider, for instance, has a sweet taste, but when it begins to ferment, little bubbles of gas appear; it loses its sweetness, and takes a sharp, biting taste. The sugar in the cider has changed into two other substances, the gas (carbonic acid) and alcohol. The formation of the gas causes the bubbling and sparkling; the alcohol gives the pungent taste. Liquids containing starch, obtained by soaking rye, corn, barley, and other grains, can be made to undergo fermentation in nearly the same way. The starchy matter changes into sugar, and the sugar into carbonic acid and alcohol. Fermentation then is the change of organic substances by which their starch, sugar, etc., are decomposed, under the influence of water, air, and warmth, and their elements recombined into new compounds.

What is distillation?

Distillation is the changing of a liquid into a vapor by the application of heat, and then condensing the vapor into a liquid again by cooling it. Some liquids boil, or change to vapor, at lower temperatures than others; and this fact is made use of in order to separate one liquid from another by distillation, as alcohol from water. Alcohol becomes vapor when heated to 173° F., water when heated to 212° F.

How are alcoholic liquors obtained?

Alcohol is produced in only one way, by the fermentation of some liquid containing sugar or starch. All the so-called alcoholic liquors, brandy, rum, gin, pure alcohol, etc., are separated from these fermented liquids by distillation. The fermented liquid is heated to a temperature sufficient to change the alcohol into vapor. This vapor is collected in a cool receiver and condensed. The result is a new liquid containing a much larger per cent. of alcohol.

It is not pure alcohol, because some steam and fragrant vegetable substances pass into the cool receiver along with the alcoholic vapor. The taste, or flavor, of the distilled liquor will vary just as the taste of the fruit or vegetable from which the fermented liquid is obtained varies. It is in this way that each kind of liquor obtains its characteristic flavor and odor.

Brandy is distilled from the fermented juice of grapes, or wine; rum is obtained in the same way from fermented molasses; gin from fermented barley and rye, and afterwards flavored with juniper berries; whiskey from fermented corn, rye, barley, or potatoes; the alcohol of commerce is obtained by re-distilling whiskey. These distilled liquors contain from forty to fifty per cent. of alcohol; the rest is water flavored according to the vegetable substance used in their manufacture.

Beer, wine, and cider are not distilled liquors, and contain a much smaller amount of alcohol.

Mention some good uses to which alcohol may be put.

Alcohol has many valuable uses in medicine and the arts. Many medicines are prepared by mixing drugs with it. It is used for dissolving gums and resins used in paints and varnishes. Perfumes and extracts for flavoring, etc., are made by combining it with various oils and essences. It is useful as a burning-fluid where great heat and little light are desired. As it will not freeze, it is used in making thermometers. It absorbs the water, and so prevents the decay of substances put into it; it is therefore useful for preserving the bodies of small animals, insects, and other specimens for museums, etc. Many others might be mentioned, but these are sufficient to show that alcohol, rightly used, is a valuable servant.

CHAPTER XXV.

STIMULANTS AND NARCOTICS.—CONTINUED.

Is alcohol a food?

Alcohol is composed of carbon, oxygen, and hydrogen only. It does not contain nitrogen, therefore it lacks one important element, and it will not make muscle. It contains no iron for the blood, and no phosphorus or lime for the bones. It leads to the overgrowth of some tissues, as connective tissue and fat; but it does not develop bone, muscle, or brain. It does not satisfy hunger except by dulling the sensibility of the nerves of the stomach. A true food becomes a part of the tissues of the body, but alcohol appears to remain in the body unchanged, until it is cast out in vapor from the lungs, pores of the skin, and by the kidneys. The conclusion is that alcohol is not a true food.

Does alcohol help to keep up the warmth of the body?

Many people take alcoholic drinks under the impression that they afford warmth, but the increased bodily temperature which they produce is only temporary and is due mainly to the stimulating effects of the alcohol upon the circulation. This slight increase, however, is more apparent than real; because, in fact, the stimulated circulation has brought more blood to the surface, and so increased the heat of the surface at the expense of the inner portions of the body. This surface heat is rapidly lost by radiation, so that in the end the effect has been to diminish the warmth of the body, instead of increasing it. Tests made by army officers, Arctic explorers, and others who have given special study to the question, go to prove that the use of alcohol lessens the power to endure extremes of heat or cold.

Is the moderate use of alcoholic drinks safe or beneficial?

In almost every instance the moderate use of alcoholic drinks, as in beer-drinking, etc., is one of the stages which brings the drinker to their immoderate use in stronger forms and in increased quantities. An eminent physician says that it is a physiological fact that the attraction of alcohol for itself is cumulative; that so long as it is present in the human body, even in small quantities, the longing for it, the sense of requirement for it, is present; and that, as the amount of it insidiously increases, so does the desire. There is an abundance of emphatic testimony which goes to prove that the use of alcohol, to an extent far short of what is necessary to produce intoxication, injures the body and diminishes mental power. The drinker himself may not be conscious of these effects, but those who know him best can see that it takes something from the fineness of his character, interferes with the keenness of his judgment and the evenness of his temper. Absolute security is found only in total abstinence.

Is the taste for alcohol inherited?

In answer to this question Dr. Willard Parker of New York says: "There is a marked tendency in nature to transmit all diseased conditions. Thus, the children of consumptive parents are apt to be consumptive. But, of all agents, alcohol is the most potent in establishing a heredity that exhibits itself in the destruction of body and mind. There is not only a propensity transmitted, but an actual disease of the nervous system."

Dr. Albert Day says: "Were it now possible to deprive every man, woman, and child of intoxicants in the future, I believe it would require a century to eliminate by the natural laws of evolution the disease produced in the past, by alcoholic indulgence." Beyond a doubt the effects of alcohol are transmitted from parent to child, from generation to generation, in a tendency to nervous diseases, mental weakness, imbecility, and insanity.

Does alcohol cause a tendency to disease?

There is the best of medical authority for saying that alcohol alters and impairs tissues so that they are more prone to disease. It causes an increased liability to fevers and inflammations. Yellow fever is more apt to be fatal when it attacks those who use liquors freely; the same thing is true with persons who have pneumonia, or inflammation of the lungs. Life-insurance tables show that a temperate man stands a better chance of a long life than an intemperate man. According to such tables, at the age of twenty, the temperate man has a chance of living 44.2 years; at the age of thirty his chance is 36.5 years; at forty it is 28.8 years. At the age of twenty the intemperate man's chances are 15.6 years; at thirty, 13.8 years; at forty, 11.6 years. This is a gain of twenty years or more in favor of the temperate man.

Does alcohol weaken the moral character?

One of the first effects to be noticed, after indulgence in alcoholic drinks, is the lack of self-control. The half-drunken man is pleased or enraged without sufficient cause; he says and does rash things, his logic is muddled, he is unable to appreciate nice shades of right and wrong, and with this general mental weakness comes a dulness of the moral sense. The newspapers abound in accounts of thefts, assaults, riots, and murders, committed by persons under the influence of intoxicating liquors, and prisons are full of men undergoing punishment for crimes committed under such circumstances.

Describe the stages of intoxication by alcohol.

The effects of alcohol upon the nervous system may, according to Dr. Richardson, be divided into four successive stages, as follows : first, the stage of excitement ; second, the stage of muscular weakness ; third, the stage of mental weakness ; fourth, the stage of unconsciousness.

In the *first stage,* the nerves which lead to the capillaries and small blood-vessels become partially paralyzed, so that they are unable to regulate the flow of blood. The restraint being taken away, the heart beats faster, and the blood rushes through the system with increased force. The first effect is usually a feeling of animation and good nature. The increased flow of blood stimulates the brain and muscles to greater activity, but it confers no permanent power to either. The extra work which a man is able to do under such circumstances is only temporary, and will be followed by a corresponding weakness ; if talk or conversation has been brilliant, it has been made so at the expense of the judgment and understanding.

As the influence of the alcohol advances to the *second stage,* it begins to produce its narcotic effects. The spinal cord is affected, and the drinker loses control of some of his muscles. The muscles of the legs, lower lip, and eyelids seem to fail first. This appears in the unsteady gait and in the leering, expressionless look of the drunkard, caused by a lack of tension of the muscles about the mouth and eyes.

In the *third stage,* or stage of mental weakness, the stimulating effect has begun to pass off. The memory has begun to fail, the will and the judgment are impaired, and all the powers of the mind are in confusion. The tongue is loosened, and self-control nearly or altogether lost. The lower, or the animal nature, takes control of the man, and education, self-respect, and social restraints have lost their power over him. He is quick to take offence at real or fancied affronts, and is liable to commit crimes and do deeds of violence at the slightest provocation.

The *fourth stage* is the stage of stupor, or unconsciousness. The narcotic properties of the alcohol have completed their work, and, in common speech, the man is "dead drunk," motionless and insensible. And were it not for the fact that the nerve-centres which control the action of the heart and organs of breathing are the last to be affected, every case of complete intoxication would be a case of death. This condition remains for an indefinite number of hours, when gradually the mind rallies from the stupor. For several days there is a distressing weariness of nerves and muscles, which shows how powerful were the effects of the alcohol, and with what difficulty the injured organs recover their normal condition. And it is doubtful whether a brain once thoroughly intoxicated ever fully recovers from the effects.

What is the nature of delirium tremens?

Delirium tremens is a nervous disease due to the excessive use of alcohol. It is a frightful form of temporary insanity. It may be caused by a single fit of intoxication, but it is usually the result of an excessive and long-continued use of intoxicating liquors. The victim is nervous and restless; he sees spectres, usually foul and horrible, about him; he imagines that he is surrounded by snakes and frightful monsters, from which he tries to escape. The delirium may continue until the victim dies from exhaustion, or until he sinks into a stupor, from which he may awaken comparatively sensible; but unless the strength of the victim has been greatly reduced by the use of alcohol, delirium tremens is not usually fatal.

www.ingramcontent.com/pod-product-compliance
Lightning Source LLC
Chambersburg PA
CBHW030402170426
43202CB00010B/1458